Recent Trends in
Biological Pest Control

The Editor

Dr. Sathe Tukaram Vithalrao [M.Sc., Ph.D., Sangit Vishard, IBT (Seri.), F.I.S.E.C., F.S.E.Sc., F.S.L.Sc., F.I.C.C.B., F.S.S.I.] is presently working as Professor and Head, Department of Zoology, Shivaji University, Kolhapur. He has teaching experience of 29 years in Entomology at University PG department and 15 years in Agrochemicals and Pest Management. He has written 30 books and published 255 research papers in national and international journals of repute. He guided 20 Ph.D. students and completed 6 major research projects (from CSIR, DST, DBT and UGC). He visited Canada (1988), Japan (1988), Thailand (2002, 2004), Spain (2005), France (2005), South Korea (2006) and Nepal (2007) etc. for academic work. He is member of editorial board of eleven prestigious journals. He delivered 35 talks through All India Radio and internal conferences and involved in Doordarshan, S.T.V. and B. T.V. programmes on useful and harmful insects. He published more than 35 popular articles in daily newspapers on insects and sericulture. He got several prestigious awards like "Environmentalists of the Year-2003", "Bharat Jyoti", "Jewel of India", "International Gold Star", "Eminent Citizen of India", "Education Acumen", "Best Educationist", "Eminent Scientist of the Year-2008", "Lifetime Education Achievement", "Lifetime Achievement in Entomology and Insect Taxonomy-2009", Educational Leadership-2011, Asia Pacific International Award-2012, Global Education Leadership Award-2013, etc. He is also working as Research and Recognition (RR) Committee member for Pune University, Pune; North Maharashtra University, Jalgaon; Shivaji University, Kolhapur and DBA Marathwada University, Aurangabad. He has been awarded several fellowships from different scientific and academic societies. He is Chairman of Maharashtra District Environmental Centre of NESA.

Recent Trends in Biological Pest Control

— Editor —

Prof. (Dr.) T.V. Sathe

Head
Department of Zoology
Shivaji University
Kolhapur – 416 004, M.S.

2014

Daya Publishing House®

A Division of

Astral International Pvt. Ltd.

New Delhi – 110 002

Published by   : **Daya Publishing House®**
                        A Division of
                        Astral International Pvt. Ltd.
                        – ISO 9001:2008 Certified Company –
                        4760-61/23, Ansari Road, Darya Ganj
                        New Delhi-110 002
                        Ph. 011-43549197, 23278134
                        E-mail: info@astralint.com
                        Website: www.astralint.com

Laser Typesetting   : **Classic Computer Services**, Delhi - 110 035

Printed at   : **Replika Press Pvt. Ltd.**

PRINTED IN INDIA

PREFACE

Biological Pest Control has attracted the attention of farmers and scientists at global scenario since this method is eco-friendly, pollution free and provides permanent solution for pest control. The book contain recent and different trends of biological control of insect pests, such as the use of braconids, Ichneumonids, chalids, hemipterans, carabids, lady bird beetles, lace wings, grasshoppers, tachinids and weevils and use of vertebrate biocontrol agents like pisces, amphibians and birds in pest control. Emphasis is given on the fundamental and basic aspects of biological control of insects as well as on biological control of sugarcane and mulberry pests and mosquitoes. Emphasis is also given on aspect of weed control by insects. Thus, the book provides ready work for biological pest control and references for research workers. No such diverse work is available in the form of book on biological control of insect pest. This is unique attempt. Hence, book will be helpful and stimulatory to students, teachers, scientists and farmers in the field of pest management. The editor is thankful to the contributory authors and workers whose work is cited in the text. Prof. T.V. Sathe is thankful to DBT, DST, UGC and CSIR for there continuous financial assistance for his research work and to Madhuri Sathe and Dr. Nishad Sathe for their help in many ways in completion of this work.

Prof. (Dr.) T.V. Sathe

CONTENTS

Chapter 1

BIOLOGICAL CONTROL OF INSECT PESTS

*T.V. Sathe**

Department of Zoology, Shivaji University,
Kolhapur – 416 004, Maharashtra, India

India is Agricultural country. Agriculture plays an important role in economy of India. About 150 million hectares of land is under agricultural cultivation. India cultivate varieties of agricultural crops such as cereals, pulses, vegetables, oil seeds, fibers, fruits, narcotic and spices etc. Cash crops like sugarcane, grapes, cotton etc. are also widely cultivated in India. However, expected yield of crops is not achieved so far because of insect damage. Within the insect group harmful insects are about 2 per cent. Mosquitoes, Bed bug, Rat flea, housefly, Sandfly etc. are the household insect pests. They have important role in transmitting several diseases. Aphids, Jassids, mealy bugs, stem borers, beetles, thrips, termites etc. cause damage to agricultural crops. Insect damage to our property is due to their basic needs such as food, shelter and mate. Recently, India faced the problem of rat flea. Every year insects damage our property costing more than Rs. 500 Crores.

Indian Agriculture should be made more modern. Till we are reliable on pesticidal use in insect pest management. In 1950 the production of pesticides was 4000 metric tons and is increasing every year by 100 metric tones.

The widely use of pesticides in insect pest management has resulted several serious problems such as:

* E-mail: profdrtvsathe@rediffmail.com

1. Air, water, soil pollution
2. Health hazards
3. Killing of beneficial insects and non target organisms
4. Pest resistance against pesticides
5. Pest resurgence
6. Secondary pest out break etc.

The above things clearly indicate that there is need of an alternative for pesticides. Biological pest control is very good alternative for pesticidal control of insects. Infact, biological control is "living weapon" over chemical control. Therefore, biological control is accepted through out the world as ecofriendly method of insect pest management.

What is Biological Control?

Biological control is the suppression of population of target species below the level of economic damage by using parasitoids and predators. In old days, pathogens were also considered as biocontrol agents. But recently, pathogens are studied under microbial control.

Historical Account

The first parasitoid *Cotesia* (*Apanteles*) *glomeratus* Cameron has been reported in 1602 killing cabbage caterpillar *Pieris rapae*. Arab farmer first used predatory ants to control citrus caterpillar in 1775. A sugarcane pest, Red locust *Nomadacris septemfasciata* has been controlled by using Indian mynah bird *Acridotheres tristis* in Mauritius as first International movement of biological pest control in 1762. Later, in 1860 the cottony cushion scale *Icerya purchasi* was controlled by using a predatory lady bird beetle *Rodalia cardinalis* and a parasitoid *Cryptochaetum icerye* in California, USA. Then several biocontrol agents have been used in pest management (Sathe and Bhoje, 2000).

Ecological Basis

Biological control has ecological basis. As like temperature, humidity, light, rainfall etc parasitoids and predators are population regulatory factors in ecology. In biological control programme equilibrium of two populations *i.e.* pest and biocontrol agent is to be maintained. Hence, it is said that biological control has ecological basis.

Economics

A very large amount is spend on chemical control although we couldn't get permanent pest control but, by spending very less amount on biological control one can achieve permanent control of pests. California industry saved 115 million dollors by spending only 4-5 million on biocontrol programme in 1923. In general, biological control provides handsome dividend to farmers. The Table 1.1 shows economical details of some biocontrol programmes.

Table 1.1: Economics of Biocontrol of some Insect Pests.

Sl.No.	Natural Enemy Used (species)	Dose	Pest Species	Required Cost
1.	*Trichogramma* spp.	40,000/acre	*Chilo* spp.	80.00
2.	*Trichogramma* spp.	1,00,000/acre	*Erias* spp. *Pectinophora gossypiella* *Helicoverpa* sp.	150.00
3.	*Bracon brevicornis*	1800/acre	Coconut, Black headed caterpillar	150.00
4.	*Cryptolaemus montrouzieri*	600/acre	Mealy bugs	350.00
5.	*Scymnus* spp.	600/acre	Mealy bugs	350.00
6.	*Goniozus nephantidis*	1200/acre	Coconut, Black headed caterpillar	200.00

Applications

The biological control can be applied for control of insect pests of agricultural crops, horticultural crops, forest crops and pests of medical and veterinary importance etc. Biological control is equally effective against weeds.

Steps

Biological control of pest species is not simple task. It can be achieved by adopting following steps.

1. Correct identification of pests and biocontrol agents (BCAS)
2. Study of origin, geographic distribution and ecological requirement of pests and BCAS.
3. Study of check list of pests and BCAS.
4. Prediction of success of efficacy of BCAS.
5. Collection and preservation of BCAS
6. Shipment of BCAS
7. Quarantine measures
8. Propogation of BCAS and pest species.
9. Release of BCAS and colonization
10. Follow up recoveries
11. Re-evaluation of programme.

Methods

There are four methods of biological control.

1. Use of labour for collection of pests and their killing.
2. Collection of BCAS from natural habitat and releasing them into target area.

3. Mass rearing of BCAS and their periodic release.

4. Importation of BCAS, mass production and release.

Future of Biological Control

It is crystal clear that pesticides will never solve the permanent problem of pests but, biological control when works, can solve the permanent problem. Chemical control leads several serious problems to human beings and nature. International societies/agencies like FAO, WHO, CIBC, IOBC etc. are coming forward to encourage workers in biological pest control by providing technical information and financial assistance. Biological pest control is ecofriendly. Hence, it has bright future.

Biocontrol Agents

1) Parasitoids (Figures 1.4–1.6, 1.8–1.10)

The parasitoids are entomophagus insects and are different from true parasites. The parasitoids lay their eggs in/on the host body and develop upon them and then kill the host at the end of its association. Parasitoids are scattered in several insect orders like Hymenoptera, Diptera, Lepidoptera, Hemiptera etc. However, Hymenoptera ranks first as far as the number of species is concerned. According to Kerrich (1960) the existing number of parasitic hymenoptera is about 2,50,000 in the world. The most important families of parasitic Hymenoptera are Ichneumonidae, Braconidae, Trichogrammatidae and Chalcideae. 60,000 species of Ichneumonids and 40,000 species of Braconids have been described in the world (Gupta, 1988). Several species of Hymenoptera attack the pests of economic importance (Sathe, 1994). *Isotima javensis* attacks sugarcane top borer (*Tryporyza nivella*), *Xanthopimpla punctata* (Figure 1.1 and 1.3) to Jowar stem borer (*Chilo partellus*) (Figure 1.2), *Apanteles* spp. (Figure 1.10) to several lepidopterous and homopterous pests. In addition, Encyrtids, Eulophids, etc are also potential biocontrol agents of several insect pests. However, the Trichogrammatids specially, *Trichogramma* spp. are very prominently figured in the biological control programmes of the world. The most significant characteristics of this genus is that they are very small in size and attack very first stage (egg stage) of the pest and check the future damage by pest to crops and thus, the population of the target species. The important species of *Trichogramma* (Figure 6.2) used in biological pest suppression are *T. minutum, T. chilonis, T. evanescens, T. australicum, T. brasilensis, T. confusum, T. japonicum, T. armigera, T. orstriniae, T. acheae* etc.

From order Diptera the family Tachinidae ranks first. More than 1500 parasitic species of tachindflies are known to science (Askew, 1971). They parasitize several lepidopterous and coleopterous pests of economic importance.

2) Predators (Figures 1.11 and 1.12)

As like parasitoids, predators have also a very significant role in biological pest control. The predators are generally larger than their preys. They consume several preys (pests) in their life by hunting. Hence, they are more efficient than the parasitoids. The predators are widely scattered in several groups of animals like birds, reptiles, amphibians, mammals, pisces, coelentrates, arachnids (spiders), insects, etc.

Class Insecta

The orders of class insecta like Coleoptera, Neuroptera, Hemiptera and Diptera provide a large number of predacious species. From Coleoptera, the family Coccinellidae along gives 4000 predacious species and commonly called as Lady bird beetles (Coccinellid beetles) (Figure 1.11). The coccinellid, *Menochilus sexmaculata* (Figure 1.2) feed on many homopterous pests like aphids, mealy bugs, aleyrodids, etc. Likely *Coccinella* spp. also feed on aphids. The tiger beetle, *Pheropsophus sobrinus* predates on the larvae of rhinocerous beetle, *Oryctes rhinoceros*. The hemipteran ruduviid, *Harpactor costalis* feed on red cotton bug, *Dysdercus* sp; Mirids, *Psollus* spp. on thrips while, the pentatomid bugs *Cantheconidia* sp. and *Andrallus* sp. feed on the larvae of *Helicoverpa* (*Heliothis*) *armigera* (Figure 1.7) and *Spodoptera litura* (Figure 16.9). The families Heliodinidae, Lycaenidae and Noctuidae of order Lepidoptera also contain some predaceous species. The Heliodinid (*Stathmopoda* sp.) feed on some coccids.

As like Coleoptera, Neuroptera order has also great potential as predatory species and hence utilized prominently in biocontrol of pest species. Lace wings and ant lions are prominent predators from this order. The praying mantids (Figure 1.2) belongs to order Dictyoptera have also tremendous potential in suppression of pest populations. *Mantis religiosa* predates on several types of insect pests including Hemiptera and Lepidoptera etc.

Invertebrate Predators

The spider, *Xystichus* spp. predate on several species of sawflies. Many species of mites are predaceous on insects belonging to more than 10 orders. The Coelentrates (hydra) *Chlorohydra* spp. are effective against mosquitoes, *Aedes higromaculis* while, Planaria, *Dugesia dorotocephala* is effective against *Culex* mosquitoes.

Vertebrate Predators

Many small species of pisces are insectivorous. They feed on larvae of mosquitoes and biting flies. The amphibians are practically mainly insectivorous. The toads *Bufo* spp are very good biocontrol agents of insects. *Bufo marinus* was very successfully introduced in Puerto Rico in 1920 for controlling Junebeetles, mole crikets and cockroaches. The giant toad feed on white grubs.

Birds

The birds have very significant role in biological pest control. Wood-peckers, wood warblers and certain migratory birds feed on black beetles, spruce bud worms and pine saw flies respectively. The Indian mynah bird, *Acridotheres tristis* is classic example of biological control agent from birds. The Indian crows (grey striped) are found feeding very potentially on *Helicoverpa* (*Heliothis*) *armigera* larvae.

Mammals

The bats (Chiroptera) feed on several nocturnal flying insect pests including locusts and certain moths. Moles and shrews are largely insectivorous. The Indian false vampire, *Megaderma lyra* feed on tasar silkmoths and grasshoppers. The striped squirrel, *Funambulus pennanti* is effective biological control agent of crikets, *Gryllotalpa* sp., locusts and hoppers. Likely, foxes, jackals, wild dogs, domestic dogs, mongoose, cats etc. are also good biological control agents of insects.

Present Status of Biological Control

Since biological control is hazard free and pollution free, it is a very good alternative for chemical control. In USA there are 50 commercial insectaries, actively engaged in mass rearing of various parasitoids and predator species. 3000 million *Trichogramma* spp. 203 million *Aphytis melinus*, 26 million *Cryptolaemus montrouizeri* and 18 million *Crysoperla carnea* are mass reared for the control of various lepidopterous pests, scales, mealy bugs and other soft bodie sucking insects.

In Rusia more than 10 biological factories are producing about 50,000 million *Trichogramma* spp. per season. Besides, about 15 species of parasitoids and predators are mass reared and released in about 10 million hectares, for controlling a variety of pests in Rusia. Several species of *Trichogramma* are mass reared in China and used against a variety of crop pests of which cotton alone comprised about 6,80,000 ha annually. In Britain and Netherlands 170 million *Encarsia* spp. and 32 million *Opius* spp are reared for the control of white flies and olive flies respectively. Several natural enemies have been imported from different countries to control insect pests in India. Some of them are listed in Table 1.2. In India also several species of biocontrol agents are mass reared and utilized in biological pest suppression. (Rao *et.al.* 1971; Sathe, 2004).

In India, some major pests like sugarcane borers, pyrilla, cotton bollworms, rice stem borer, coconut black headed caterpillar, sugarcane scales and several species of mealy bugs have been controlled by biological means (by using insects). The mass rearing techniques have been developed for many natural enemies of insect pests notable examples are *Trichogramma* spp. *Chrysopa* sp. Lady bird, beetles etc. There are several potential natural enemies which are not fully exploited in pest control in India. The commercial production of biocontrol agents was started in 1981 in India and now a days, several laboratories are actively engaged in producing biocontrol agents on large scale. The natural enemies presently commercially available include various species of *Trichogramma* for controlling sugarcane borers, cotton boll worms, codling moth, *Helicoverpa armigera* etc. and *Chilonus blackburnii, Cryptolaemus montrouzieri, Crysopa* sp. etc. for several other pests (Table 1.2).

Application Dose

The recommended doses of natural enemies against the pest species are shown in Table 1.3.

Disadvantages of Biological Control

1. Take long time to control pests.
2. It is not suitable for the short harvesting crops.
3. It has slow pace.
4. It is not applicable to all species of the pests.

Table 1.2: List of some Biocontrol Programmes in India.

Sl.No.	Year and Country (Imported)	Natural Enemy	Pest Species	Crop	Established Place
1.	1929 Australia	*Rodalia cardinalis* (Coleoptera)	*Icerya purchasi* (Hemiptera)	Citrus	T.N., Kerala
2.	1940 USA	*Aphelinus mali* (Hymenoptera)	*Erisoma lanigerum* (Hemiptera)	Apple	Punjab, Assam, S.India
3.	1960 Zanzibar via Sri Lanka	*Spoggosia beniziana* (Diptera)	*Nephantis serinopa* (Lepidoptera)	Coconut	T.N., Orissa
4.	1964 New Guinea	*Telenomus* sp. (Hymenoptera)	*Achea janata* (Lepidoptera)	Caster	A.P.
5.	1965 S.U.S.A.	*Trichogramma brasilensis* (Hymenoptera)	*Pectinophora gossypiella* (Lepidoptera)	Cotton	Haryana
6.	1968 S.U.S.A.(California)	*T. brasilensis* (Hymenoptera)	*Helicoverpa armigera* (Lepidoptera)	Polyphagus	Karnataka
7.	1973 Mauritus via East Africa	*Sticholotis medayassa* (Coleoptera)	*Melanospis glomerata* (Hemiptera)	Sugarcane	India

Figures 1.1–1.12: Biocontrol Agents of Insect Pests.

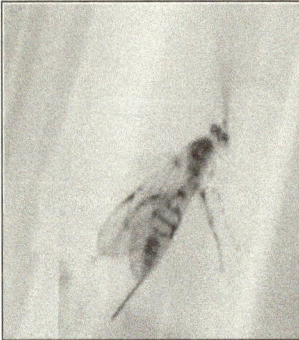

Figure 1.1: *Xanthopimpla*

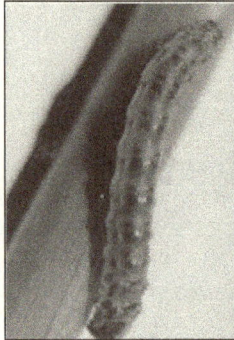

Figure 1.2: *Chillo partelus*

Figure 1.3: *Xanthopimpla*

Figure 1.4: *Enicospilus*

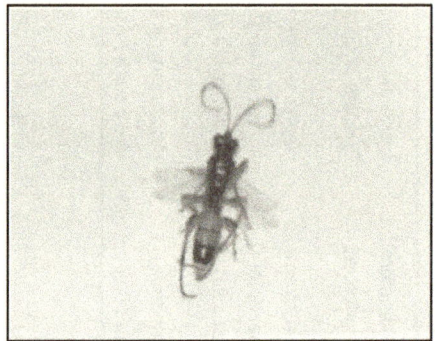

Figure 1.5: *Goryphus*

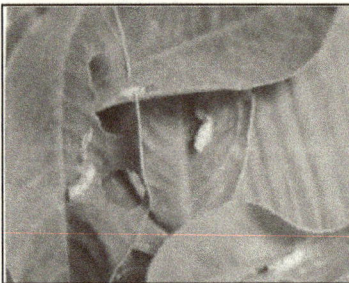

Figure 1.6: *Campolitis* Cocoons

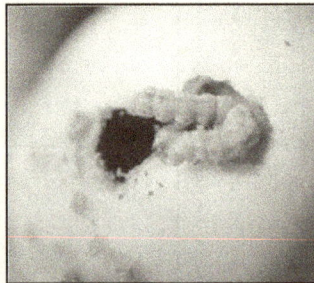

Figure 1.7: *H. armigera*

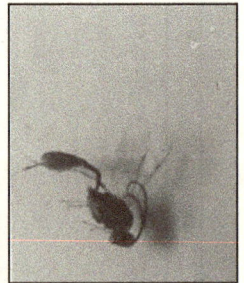

Figure 1.8: *Campolitis*

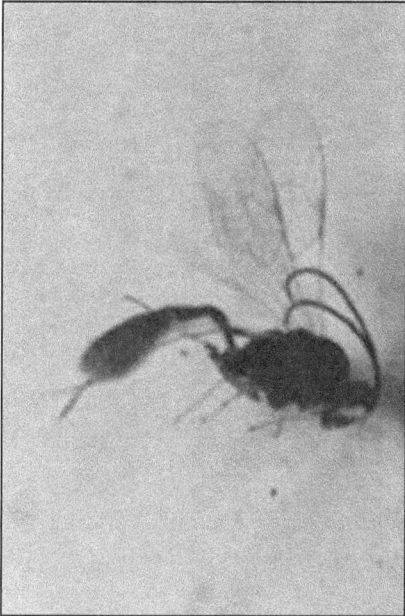

Figure 1.9: *C. chlorideae*

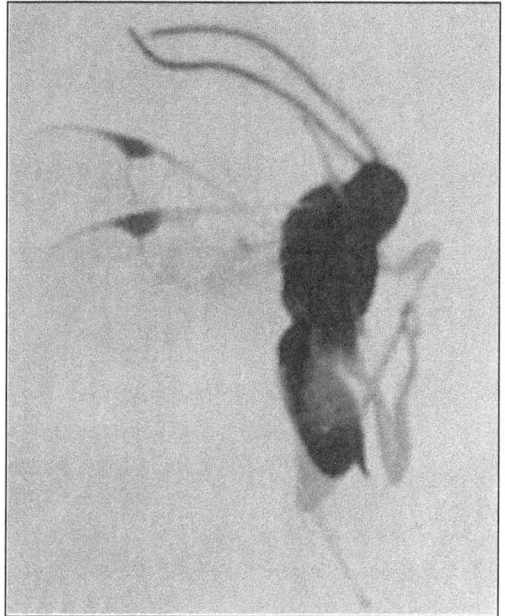

Figure 1.10: *Apanteles*

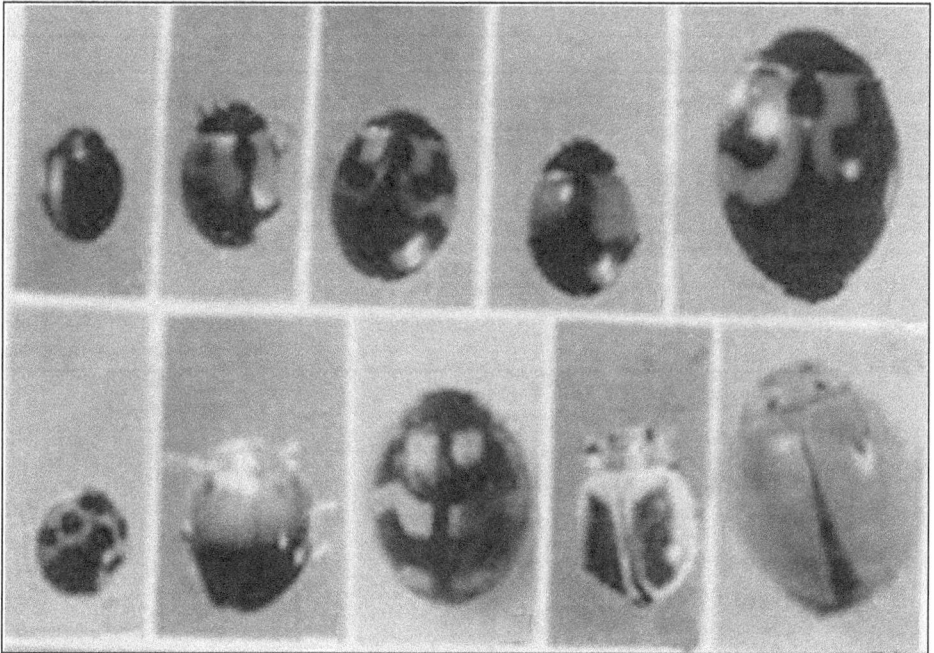

Figure 1.11: Ladybird Beetles

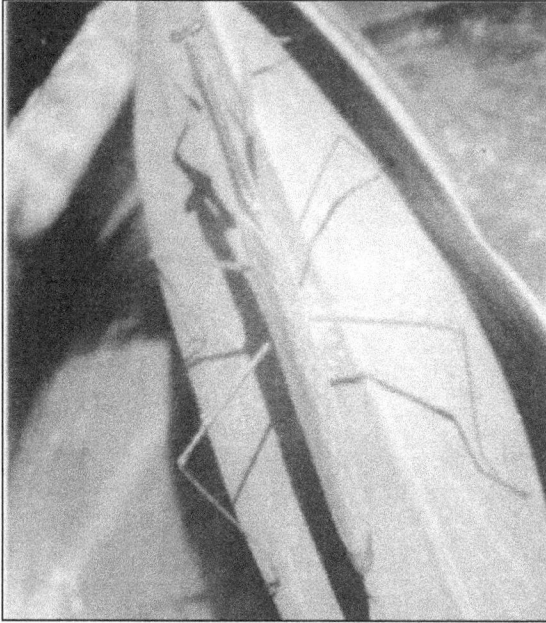

Figure 1.12: Praying Mantid

Table 1.3: Biocontrol Agents Available and Recommended Doses in India.

Sl.No.	Species Natural Enemy	Pest	Dose Recommended
1.	*Trichogramma* sp.	*H. armigera*	120,000/acre
2.	*Trichogramma* sp.	*Chilo sacchariphaga indicus*	2,00,000/ha
3.	*Chilonus blackburnii*	*Cotton bollworms*	10,000/week/acre
4.	*Bracon krikpatrikii*	*Cotton bollworms*	5,000/week/acre
5.	*Meteorus dichomeridis*	*Spilosoma obliqua*	1,00,000/ha/week
6.	*Cryptolaemus montrouzieri*	*Maconellicoccus hirsutus*	@250/adults/ha
7.	*Nesolynx thymus*	*Exorista bombycis*	1,00,000 adults females and 500 males in 3 split releases.

Advantages

1. It is pollution free and hazard free.
2. It solves the permanent problem of pests when works.
3. It is economic method.
4. It is ecofriendly method.

Chapter 2

BIOLOGICAL CONTROL
OF MOSQUITOES

T.V. Sathe*

*Department of Zoology, Shivaji University,
Kolhapur – 416 004, Maharashtra, India*

Introduction

India is considered to be one of the richest centers of biodiversity in the world, particularly because of the large number of ecosystems. Mosquitoes are diversified taxonomical group of insects. Taxonomical biodiversity of mosquitoes is an important aspect of medical science. Mosquitoes are responsible for causing dreaded diseases to human beings. Malaria, Filariasis, J.E., Chikunguniya, Dengue etc. are the common diseases caused by mosquitoes, which are fatal to human being. According to Stone *et al.* (1992) there are about 2700 species of mosquitoes in the world. Revised estimate of mosquitoes is 3500 described species (Sathe and Tingare, 2009). *Anopheles, Culex* and *Aedes* are important genera of mosquitoes of medical importance. Malaria is caused by *Anopheles* species. Filariasis is caused by *Culex* species while, Dengue is caused by *Aedes* species.

Mosquitoes are characterized by small size, 3.00 to 6.00 mm in body length, delicate and slender body, covered with hair and scales. They are black or brown, often spotted white and have piercing and sucking type of mouth parts for sucking the blood of animals or cell sap of plants. The larvae of mosquitoes are elongate wrigglers and aquatic in habitat. The pupae are also aquatic and capable of swimming

* E-mail: profdrtvsathe@rediffmail.com

by paddel like movements of the abdomen. The larvae breed in all kinds of fresh water and in the brackish water. The larvae feed mostly on minute algae and other organic matter floating in the water.

The control of mosquitoes through pesticide is difficult task due to the development of resistance in mosquitoes. Secondly, pesticides lead to serious problems such as air, water and soil pollution; killing of beneficial organisms, health hazards, secondary pest - outbreak, pest resurgence, destruction of ecocycles, etc. Therefore, ecofriendly mosquito control technique is the need of the day. Biological control is potential alternative and living weapon over chemical control. In the present topic various biocontrol agents have been discussed with their diversity and potential in control strategies.

Biological Control of Mosquitoes through Pisces

In past, several workers (Brown, 1927; Sweetman, 1958; Kalra *et al*, 1967; Greatheard, 1971; Rao *et al.*, 1971; Siogren, 1972; Legner *et al.*, 1974; 1975 a,b; Sathe and Bhoje 2000 and Sathe and Girhe 2002 etc.) have attempted the work related to control of mosquitoes using Pisces. Several insectivorous fishes are scattered in tropical and temperate regions in both fresh and brackish waters. In biological pest control of mosquitoes, the most important, effective and widely utilized fish species belong primarily to the family poeciliidae and to a lesser degree to the cyprionodontidae. Sweetman (1958) listed mosquito controlling species of Pisces.

According to Greatheard (1971) laboratory experiments in 1972 in SeychellesIslands gave strong indications of the value of *Pachypanchax playfairi* (Gunther) as a predator of Mosquitoes. Field trials were not taken in the above region. However, the fish species gave very encouraging results for mosquito control in wells and tanks with clear water in East Africa (Zanzibar).

Gambusia affinis Baird and Girard and *Poecilia reticulata* Peters are commonly used fishes in anti mosquito work. *G. affinis* is commonly called the 'mosquito fish' which reaches to breeding size within 2-3 months after its birth. It has maximum of 200 progeny per brood. The number of broods per year varies with the localities, but 4-6 broods annually reported.

According to Sweetman (1958) *P. reticulata* is more prolific than *Gambusia* sp. However, *G. affinis* is now considered ubiquitous, as it has been distributed and established throughout the world since early in the twentieth century.

The earliest introduction of *G. affinis* was in 1905 in Hawaii for control of mosquitoes (Legner *et al*, 1974), within two months, planning 2500 to 5000 fishes per ha achieved complete mosquito suppression. Hoy and Reed (1971) reported high-level suppression of *Culex tarsalis* Coqui by releasing 250-500 gravid females of *G. affinis* per ha in paddy field of California. While, Brown (1973) reported 60 per cent suppression of *Anopheles quadrimaculus* Say by introducing *Gambusia* in Southern USA in 1920s. *Gambusia* was an important factor for suppressing *Anopheles* population in Malagasy, particularly in open water habitats (Grethead, 1971). Likely, the genus *Epiplaytys* was also quiet potential in suppression of populations of mosquitoes in Africa (Brown, 1973). In Iran 4 species of *Anopheles* (Figure 2.4) were successfully suppressed by introducing over 1.5 million *Gambusia* in 1969. Rao *et al.* (1971) reviewed South East Asian programmes of mosquito control through introduction of *Gambusia* and Guppy.

Langer *et al.* (1975) studied the efficiency of the predatory desert pupfish. *Cyprinodon macularius* Baird and Girard on *Culex* mosquitoes. They reported that in shallow ponds of large size, desert pupfish caused mosquito breeding to cease with four weeks after their introduction.

Langer *et al.* (1975) reported that in some tropical area *P. reticulata* was preferred to *Gambusia* because of its superiority in pollution tolerance (Siogren.1972). In Srilanka *P. reticulata* has been preferred to release for successful control of *Anopheles* mosquitoes (Rao *et. al.*1971). A successful suppression of *Culex pipens fatigans* Wied in a canal carrying sullage waste was noted by Rao *et. al.*(1971) in Southern India. They also reported that the fish was well established in fresh and brackish waters of Southern India. Kalra *et. al.*(1967) studied *Lebistes reticulata* Peters with respect to mosquito control in Nagpur, India with good success.

Grethead, (1971) reviewed the successful establishment of *P. reticulata* in Senegal and successful suppression of *Anopheles* spp. in heavily infested ponds within 47 days. The above Guppy fish was successfully used in several countries for mosquito control (Coppel and Martins, 1977).

Sathe and Bhoje, (2005) recently reported that *P. reticulata* showed more preference to *Anopheles* (Figure 2.4) than *Culex* mosquitoes and has great potential for biocontrol of mosquitoes by performing 79.6 per cent predation of *Anopheles* and 58.4 per cent in *Culex* mosquitoes. However, it is crystal clear that the ideal biocontrol agent of fish should have high fecundity, short life cycle, small size, and top feeding habits, preference for mosquito larvae, adaptability and pollution tolerant capacity. Guppy shows all above features hence ideal in biological pest control programme of mosquitoes.

Biological Control of Mosquitoes through Dragonflies

According to Corbet (19999), there are two approaches to the biological control of pest namely inoculation and augmentative release (AR). Both approaches are useful for mosquito control. Under inoculative release natural enemies of pests are introduced into an environment where they are not already present, during which enemies multiply naturally until they reach a level such that they either eliminate the pest or keep the pest population down to a level of deemed acceptable to humans. However, this method has limited success. Augmentative release can be highly successful since this method entails prior estimation of the number of natural enemies needed to active suppression to required level and then relizing sufficient number in to a closed system. Augmentative release is practiced routinely in several countries for suppression of pests in green house, close environment from which neither biocontrol agent nor pest can disperse. Odonata in general are predators of several insects, houseflies and mosquitoes are prominently predated (Metcalf and Flint, 1979). Thomas *et. al.*(1988) studied predatory efficiency of nymphs of *Bradinophyga jaminata* and *Brachythemis contaminata* on mosquito larvae. They found that these Odonates were good predators of mosquito larvae.

In Cuba, Santhamarina and Mijares (1986) made field and laboratory observations to study the efficacy of nymphs of anisopterans such as *Pantala flavescens* (Figure 2.1), *Tramea abdominalis* and proved them (Figure 2.2) good predators of *Culex quinquefaciatus*. Similarly *D. frivialis* also acts as good biocontrol agent of mosquitoes.

A sizable work has been made from Japan in connection with the control of various types of mosquito larvae. Urabe *et. al.*(1986) studied and evaluated the predatory capacity and efficiency of *Sympetrum frequens* against the larvae at *Anopheles sinensis* in laboratory and reported that when the nymph size increased, subsequently the number of mosquito larvae consumption also intensified. They also reported the predator prey relationship between *Sympetrum frequens* and the larvae of *Anopheles sinensis* in rice field near Omiya. Japan has revealed that the density of mosquito larvae increased when the nymphal density of the predator become low. The distribution patterns of the predators and prey in the smaller field had a non-overlapping tendency, indicating effective predation.

A good example known about dragonflies used successfully through augmentative release (AR) is the introduction of half grown larvae of the libellylid *Czocontenemis servilia* into domestic water storage containers in Yangon (Rangoon) in Myanmar (Burma). The water storage containers were used by the aquatic stages of the yellow fever mosquito *Aedes aegypti* (Figure 2.3), which was responsible for the transmission of dengue fever in the locality. More than 92 per cent of the local population of *A. aegypti* was occupying the containers, which because of their function were easily accessible to householders and control operators. The systemic release of dragonfly larvae during the monsoon season (the time when dengue fever was being transmitted by the mosquito) rapidly depressed the mosquito populations to a level lower than that could have been achieved by any other known method including treatment by chemical insecticide. Sebastian *et. al.*(1990) demonstrated the trials of effectiveness of this approach. The Rangoon trial was very much successful, due to fact that both the adults and larvae of dragonflies were effective for controlling the population of mosquitoes as like conventional pesticide.

Bacterial Control of Mosquitoes

Two species of bacteria are largely involved in control of mosquitoes. *Bacillus thuringiensis* variety *israelensis* and B.t.strain H-14 (VCRC B17) were found to be extremely toxic to mosquitoes. B.t.strain H-14 (VCRC B17) was tested in Pondecherry region of India. The spore toxin complex of Bt was used as active ingredient. The formulation WDP at 2.5-5 kg/ha and 5-15 kg/ha were found effective in clear and polluted habitats respectively against *Culex quinquefasciatus*. Similarly, a stain VCRC B-42 of *B. sphaericus* was also found effective against mosquito larvae which was tested by the scientists, Pastour Institute, Paris France with collaboration of WHO. Pupicidal bacterial metabolite B-426 caused significant mortality in pupae and also affected adult emergence in *Culex quinquefasciatus*.

Fungicidal Control of Mosquitoes

Coelomomyces indicus, an aquatic fungus caused mortalities in larval anophelines and culicines in the field of paddy. Fungal metabolite F-24 has showed oviposition

Figures 2.1–2.4: Mosquito Controlling Dragonflies.

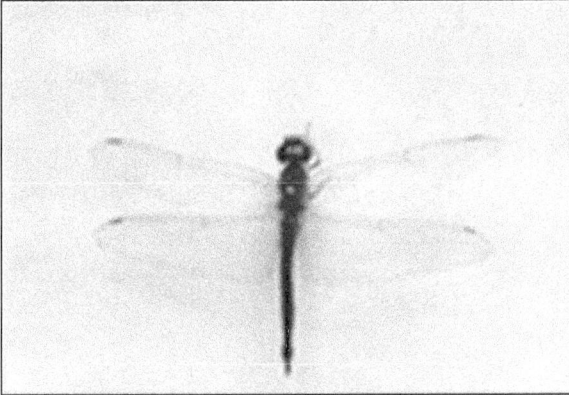

Figure 2.1 : *Pantala flavescens*

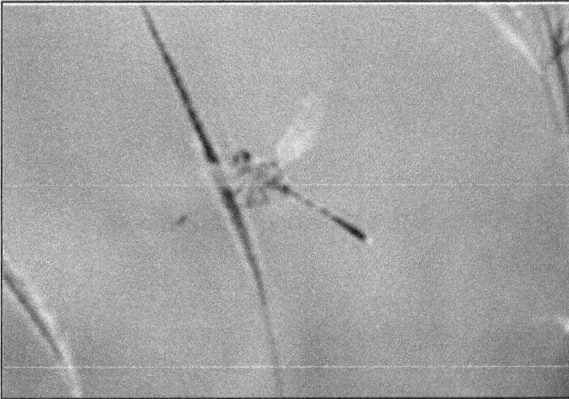

Figure 2.2: *Diplacodes trivialis*

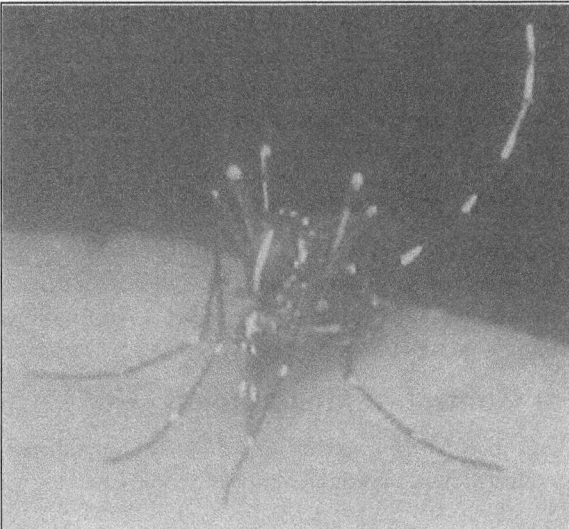

Figure 2.3: Mosquito *Aedes* sp.

Figure 2.4: Mosquito *Anopheles* sp.

attractivity against the gravid female of *Culex quinquefasciatus*. Laird, (1967) first attempted the control of mosquitoes through introduction of fungi. The programme was initiated in 1958 in TokelaIsland with release of *Coelomomyces stegomyiae* (sporangia) with the help of debris, dried bodies and sediments of tree holes against *A. albopictus* and *A. polynesiensis*. Thereafter, in 1959 survey was made which has proved the establishment of pathogenic fungi against mosquitoes. Again in resurvey 1960 native dispersal of fungus was noted and in 1963 C. *stegomyiae* parasitized up to 87 per cent larvae of mosquitoes. Mass rearing of fungi is essential but difficult. However, Roberts, (1973) attempted rearing of *C. stegomyiae* in USA.

Nematodes for Control of Mosquitoes

A mermithid nematode, *Romanomermis iyengari* has been found parasitizing aquatic stages of mosquitoes in paddy fields. The above nematode was effective against culicines breeding in semi-polluted water and Anophelines in fresh water and even *Aedes albopictus* as tree hole breeder.

Protozoans for Control of Mosquitoes

Microsporidians such *Nosema algerae* and *Amblyospora indicola* found infecting mosquito larvae in natural habitat, which lead to chronic diseases in the mosquitoes.

Hemipterans for Control of Mosquitoes

The adult as well as the nymphs of a hemipterous bug *Anisops bouveri* were effective against mosquito larvae. Similarly, *Diplonychus indicus* and *Notonecta* were also effective against *Aedes* mosquitoes.

Coelentrates for Control of Mosquitoes

Hydras are grouped under coelentrata and number of them are species of Solitary polyps occurring in fresh water bodies. Hargraves (1924) first recorded hydra as mosquito predator. The species of genera *Hydra* and *Cholorohydra* are mosquito feeders. Qureshi and Bay, (1969) studied the density and effectiveness of *Hydra americana* Hyman against the larvae of *Culex peus* Speiser. They found that this species was most effective as biocontrol agent of above species of mosquito. Legner and Medved (1972) studied naturally occurring mosquito predator hydras in California region of America. After making the survey, they mass reared *Hydra* species and released in natural habitat for control of two species of mosquitoes successfully. According to *Yu et.al.* (1974) the strain of *Chlorohydra viridissima* was mass reared and used against two mosquito species namely *Ades nigromaculis* in irrigated area and *Culex tarsalis* in river floodwater and duck club ponds. They released 500 to 1500 hydras in each habitat (One meter square net enclosure) and obtained good results, 67 per cent reduction in population of *Aedes nigromaculis* and 80 per cent reduction in *Culex tarsalis*. However, Yu *et.al.* (1974) recommends naturally occurring *Hydra* species if possible for control of mosquito species.

Planaria for Control of Mosquitoes

Plannarians in the order Tricladida are free living and mainly freshwater species. In USA, *Dugesia dorotocephala* (Woodworth) was most common in streams and ponds and readily available from most biological supply houses and was quite effective against large number of mosquitoes. Therefore, Villee *et al.* (1968) attempted various aspects of *D. dorotocephala* such as occurrence, characteristics, feeding, digestion, movement, water balance, excretion, reproduction, regeneration and polarity essential for its mass rearing. Lagner *et al.* (1975) says that *D. dorotocephala* is rearable, storable and releasable for control of mosquitoes. Legner et. al (1975) introduced *D. dorotocephala* $29/m^2$ in 1971 for the control of *Culex* sp. They observed 90 per cent reduction in mosquito population within 26 days in California. When the inoculation rate of planaria was increased in $115/m^2$, a higher-level of suppression was obtained. Planaria population in ponds was doubled in about every 30 days and Legner et.al (1975) thus, concluded that planaria were adequate substitute for chemical pesticides for mosquito control in California.

Birds for Control of Mosquitoes

A birdless country would be the most desirable place for insects. However, bird mosquito predatory prey index is not available. It is observed that Wagtail bird largely subsisting on adult mosquito diet and thus acts as good biocontrol agent of mosquitoes in Kolhapur (Sathe and Girhe, 2002). There is extreme need of preparing predatory prey index of birds and mosquitoes in India.

Conclusion

1. It seems that the programme of biological control of mosquitoes has great future since it is ecofriendly.

2. Biocontrol agents such as bacteria, fungi, planaria, hydras, dragonflies and Guppy fishes have tremendous potential in biological control of mosquitoes. Above biocontrol agents should be exploited on large scale in India since, very negligible work is known from India on biological control of mosquitoes.

3. Further study is needed on survey of species of above important groups of biocontrol agents.

4. Mass rearing of biocontrol agents need special attention for better exploitation of them in mosquito control programmes.

5. National and International funding agencies should come forward for boosting the research projects related to biological control of mosquitoes by providing funds.

6. It is extremely essential that the concept of biological control of mosquitoes should be popularized among the common people for avoiding pollution and keeping environment healthy.

References

Brown, A.W.A., 1973. Pest control strategies ten years : Malaria. *Bull. Entomol. Soc. Am.*, 19: 193–196.

Chapman, H.C. Biological control of mosquito larvae. *Ann. Rev. Entomol.*, **19**, 33–52.

Coppel, H.C. and Martins, J.W., 1977. *Biological Insect Pest Suppression*. Springer – Verlag Berlin Heidelberg. New York, pp. 119–122.

Corbet, P.S., 1999. *Dragonflies : Behaviour and Ecology of Odonates*. CornellUniversity Press, New York and Harley Books, Great Horkesty, UK, 1–829.

Greatheard, D.J., 1971. A review of biological control in the Ethiopian Region. *Commonwealth Inst. Biol. Contr. Tech. Commun.*, **5**, 1–20.

Hargreaves, E., 1924. Entomological notes from Tomato (Itlay) with reference to Facnza, during 1917 and 1918. *Bull. Entomol. Res.* **14**, 213–219.

Hoy, J.R. and Reed D.E., 1971. The efficacy of mosquito fish for the control of *Culex tarsalis* in California rice fields. *Mosq. News.*, **31**, 567–572.

Kalra, N.L., Wattal, B.L., and Raghavan, N.G.S., 1967. Occurrence of larvivorous fish *Lebistes reticulata* (Peters) breeding in sullage water at Nagpur, India. *Bull. Indian Soc. Malarial Comm. Dist.*, 4(3): 253–254.

Laird, M., 1967. A coral island experiment : A new approach to mosquito control. *WHO Chron.*, 21: 18–26.

Legner, E.F., Madved, R.A., 1972. Predator investigated for the biological control of mosquitoes and midges at the University of California. *Riverside. Proc. Calif. Mosq. Contr. Assoc.*, 40: 109–111.

Legner, E.F., Siogren, R.D. and Hall, I.M., 1974. The biological control of medically important arthropods. CRC Critical. *Rev. L. Environ. Contr.*, 4: 85–113.

Legner, E.F., Fisher, W.J., Hanser, W.J. and Medved, R.A., 1975a. Biological aquatic weed control by fish in the lower SonoranDesert of California. *Calif. Agr.*, 29: 8–10.

Legner, E.F., Medved, R.A. and Hanser W.J., 1975b. Predation by the desert pupfish, *Cyprinidon macularius* on *Culex* mosquitoes and Benthic chironomid midges. *Entomophaga*, 20: 23–30.

Metcalf, C.L. and Flint, W.P., 1979. *Destructive and Useful Insects*. Tata MaGraw Hill Publishing Company Limited, New Delhi.

Quereshi, A.H. and Bay, E.C., 1969. Some observations on *Hydra americana* Hymen as a Predator of *Culex pens* Speiser mosquito larvae. *Mosq. News* 29: 465–471.

Rao, V.P., Ghani, M.A., Sankran, T. and Mathur K.C. A review of biological control of insects and other pests in South East Asia and the Pacific Region. *Commonwealth Inst. Bil. Contr. Tech. Commun.*, 6: 54–56.

Roberts, D.W., 1975. Means for insect regulation *fungi. Ann. N.Y. Acad. Sci.*, 217: 76–84.

Santamarina and Mijares, A.,1986. Odonata as bioregulators of the larval phases of mosquitoes. *Revista – Cubana – de – Medicina –Tropical*, 38: 1, 89–97.

Sathe, T.V. and Bhoje, P.M., 2000. *Biological Pest Control*. Daya Publishing House, New Delhi, pp. 1–122.

Sathe, T.V. and Bhoje, P.M., 2005. Biocontrol potential of Guppy *Poecilia reticulata* (Peters) for mosquitoes. *Exp. Zool. India*, 8(1): 106–108.

Sathe, T.V. and Girhe B.E., 2001. Biodiversity of mosquitoes in Kolhapur district, Maharashtra. *Riv. di. Parassitologia*; 18(12): 193–198.

Sathe, T.V. and Girhe B.E., 2002. *Mosquitoes and Diseases*. Daya Publishing House, New Delhi, pp. 1–96.

Sebastian, A., Myint Myint Sein, Myat Myat Thu. and Philip S. Corbet, 1990. Supression of *Aedes aegytli* (Diptera : Culicidae) using augmentative release of larvae (Odonata : Libellulidae) with community participation in Yangon, Myanmar. *Bulletin of Entomological Research*, 80: 223–232.

Siogren, R.D., 1972. Minimum oxygen threshold of *Gambusia affinis* (Baird and Girard) and *Poecilia reticulata* Peters. *Proc.Calif. Mosq. Contr. Asso.*, 40: 124–126.

Sweetman, H.L., 1958. *The Principles of Biological Control*. W.C. Brown, Dubuque.

Thomas, M., Daniel, M.A. and Gladstone, M., 1988. Studies on the food preference in three species of *Dragonfly naiads* with particular emphasis on mosquito larvae Predation. *Bicovas*, 1: 34–41.

Urabe, V.K., Ikemoto, T. and Aida, C., 1986. Studies on Sympatrum frequency (Odonata : Libelluliadae) nymph as natural enemy of mosquito larvae, *Anopheles sinensis* in rice fields 2, Evaluation of predatory capacity and efficiency in Laboratory, *Japanese Journal of Sanitory Zoology*, 37(3): 213–220.

Villee, C.A., Walker, W.F. and J. Smith, F.E., 1968. *General Zoology*, 3rd ed. W.B.Saunders, Philadelphia – London – Tornato.

Yu, H.S., Legner, E.F. and Sjogern, R.D., 1974. Mass release effects of *Chlorohydra viridissima* (Coelenterata) on field populations of *Aedes nigromaculis* and *Culex tarsalis* in Kern County, California. *Entomophaga*, 19: 409–420.

Chapter 3

BIOLOGICAL CONTROL OF SUGARCANE PESTS

T.V. Sathe*

Department of Zoology, Shivaji University,
Kolhapur – 416 004, Maharashtra, India

Introduction

Sugarcane is cash crop of India which is cultivated in over 30 million hectares of land. Several new varieties are introduced. More than a dozen of insect pests are found attacking sugarcane crop in India. The pests cause substantial loss in crop yield and sugar recovery. Recently (2002), outbreak of sugarcane wooly aphid was experienced in Maharashtra which has paralyzed entire sugarcane industry in Maharashtra. The sugarcane is attacked by many internal feeders/borers and difficult to control with pesticides. Pest control on sugarcane requires special attention since dates of plantion can also have serious effects on pest occurrence. The crop stands for 2-3 years on a very large area which also provide favourable environment for the pests. Resistant varieties and cultural practices have their own limitations. Therefore, biological control strategies of sugarcane pests has great significance as ecofriendly method.

Natural Enemies for Sugarcane Pests

More than 200 pests are known on sugarcane. Some important species of insect pests and their natural enemies are represented in Table 3.1.

---

* E-mail: profdrtvsathe@rediffmail.com

Table 3.1: Sugarcane Pests and their Natural Enemies.

Sl.No.	Pest– Common Name	Scientific Name	Natural Enemy
1.	Sugarcane top borer	*Scirpophaga nivella* (Lepidoptera : Pyrallidae)	☆ *Isotima javensis* Rohw. (Ichneumonidae) ☆ *Telenomus beneficiens* (Zehntner) (Scelionidae) ☆ *Trichogramma chilonis* (Trichogrammatidae) ☆ *Telenomus dignoides* Nixon (Scelionidae) ☆ *Goryphus nursei* Cameron (Ichneumonidae) ☆ *Xanthopimpla pedator* Fab. (Ichneumonidae) ☆ *Rhaconotus roslinesis* Ashmead (Bebylidae) ☆ *Sturmiopsis inferens* Townsend (Tachinidae)
2.	Sugarcane shoot borer	*Chilo infuscatellus* (Lepidoptera : Pyrallidae)	☆ *Trichogramma chilonis* Ishii (Trichogrammatidae) ☆ *Trichogramma axigium* Perkins (Trichogrammatidae) ☆ *Trichogramma intermedium* How. (Trichogrammatidae) ☆ *Telenomus beneficiens* Zehntner (Scelionideae) ☆ *T. dignoides* ☆ *Cotesia flavipes* Cameron (Braconidae) ☆ *Campyloneurus mutator* Fab. (Braconidae) ☆ *Goniozus indicus* Ashmead (Bethylidae) ☆ *Sturmiopsis inferens* Townsend (Tachinidae) ☆ *Sturmiopsis semiberbis* Bezzi (Tachinidae) ☆ *Tetrastichus agyari* Rohw (Eulophidae) ☆ *Xanthopimpla* spp. (Ichneumonidae)
3.	The stalk borer	*Chilo auricilius* (Lepidoptera : pyrallidae)	☆ *Trichogramma chilonis* Ishii (Trichogrammatidae) ☆ *T. axigium* ☆ *T. intermedium* ☆ *T. chilonis* ☆ *Xanthopimpla* spp. ☆ *Cotesia flavipes* ☆ *Cotesia chilonis* ☆ *Cotesia sesamae*
4.	Sugarcane root borer	*Emmalocera depresella* (Swin.) (Lepidoptera : pyrallidae)	☆ *Trichogramma chilonis* ☆ *Cotesia flavipes* ☆ *Stenobracon* sp.
5.	Internode borer	*Chilo (Sacchariphagus) indicus* Kapoor (Lepidoptera : pyrallidae)	☆ *Trichogramma minutum*

Contd...

Table 3.1–*Contd...*

Sl.No.	Pest– Common Name	Scientific Name	Natural Enemy
6.	Pink borer	*Sesamia inferens* (Lepidoptera : Noctuidae)	☆ *Trichogramma minutum* ☆ *T. chilonis-Trichogramma* sp. ☆ *Cotesia* spp. ☆ *Xanthopimpla* spp.
7.	Pyrilla	*Pyrilla purpusilla* Walker (Hemiptera : Fulgoridae)	☆ *Epipyrops melanoleuca* Flecher (Epipyropidae) ☆ *Tetrastichus pyrillae* ☆ *Dryinus pyrillae* Kieffer ☆ *Ooencyrtus papilions* Ashmead (Encyrtidae) ☆ *Coccinella septumpunctata* (Coccinellidae) ☆ *Menochilus sexmaculatus* Fab. (Coccinellidae) ☆ *Brumus suturalis* (Fab.) (Coccinellidae) ☆ *Epiricania melanoleuca* (Lepidoptera)
8.	Mealy bug	*Saccharicoccus sacchari* (Cockerell) (Hemiptera : Coccidae)	☆ *Scymnus coccivora* ☆ *S. andrewsi* ☆ *S. nubilis* ☆ *Hyperespis triliniata* Muls. (Coleoptera : Coccinellidae)
9.	Sugarcane white fly	*Aleurolobus barodensis* (Maskell) (Hemiptera : Aleurodidae)	☆ *Encarsia issaci* Mani ☆ *E. muliyali* Mani ☆ *Eretmoceras delhiensis* L. ☆ *Azotus delhiensis* Lall.
10.	Sugarcane wooly aphid	*Ceratovacuna lanigera* Zehntner (Hemiptera : Aphididae)	☆ *Dipha* (*Conobatra*) *aphidivora* M. (Pyrallidae) ☆ *Micromus* sp. (Hemerobideae) ☆ *Chrysoperla carnea* (Neuroptera : Chrysopidae) ☆ Syrphid fly (Diptera) ☆ *Coccinella* (Coleoptera : Coccinellidae)
11.	Sugarcane white grub	*Holotrichia consanguinea* (Coleoptera : Melolonthidae)	☆ *Bacillus thuringiensis*
12.		*H. serrata*	☆ *Pheropsophus sobrinus* D. (Coleoptera) ☆ *Acridotheres tristis* (W) (Aves) ☆ *Passer domesticus* Linn.(Aves) ☆ *Corvus splendens* Viel. (Aves) ☆ Toad *Bufo melanostictus* (Amphibia)
13.		*Leucopholis lepidophora* Brum.	☆ *Anomala dorsalis* Fab. (Coleoptera) ☆ *Anomala bengalensis* (Coleoptera) ☆ *Anomala biharensis* (Coleoptera) ☆ *Heterophychus* sp. (Coleoptera) ☆ *Pentodon* spp.

Biocontrol Strategies for Sugarcane Pests

Trichogramma spp. were first used in early forties for the control of sugarcane borers, *Chilo* spp. The introduction of *Trichogramma* against borers has been resulted both positive and negative results. *Trichogramma* spp. were in use since 30-35 years. Later, in Tamil Nadu biocontrol programmes have been evaluated critically during late gernties and easy eighties and evidenced the inductive release of *Trichogramma chilonis* against sugarcane internode borer *Chilo saccariphagus indicus.* The sugarcane top borer *Scirpophaga excerptalis* Wal. has been controlled for almost permanent status with the help of an Ichneumonidfly *Isotima javensis* Rowh. Similarly, pyrilla has been controlled with the help of a Lepidopterous ectoparasite *Epiricania melanoleuca* Wal. In recent days predators have been used to control white grubs, some borers and scale insects.

In Tamil Nadu *Trichogramma chilonis* was mass reared by co-operative sugar federation. Biocontrol Research Laboratory, Chegalpattu is largely involved in biological control programme from 1982. This centre reared parasitoids covering 5000 hectares during 1986-87. 12 mg of parasitoid can yield 1,20,000 adults of Trichogramma and is sufficient dose for one acre of crop against sugarcane internode borer.

For top borer *S. nivella, T. japonicum* has been also reared and used. *Trichogramma* parasitoids are quite effective within the range of temperature from 26°C to 30°C and relative humidity 60-70 per cent.

Telenomus and *Tetrastichus* were in use against pyrilla eggs. However, for large scale production no satisfactory method was developed in early time. Hence, their use was limited. In 1965 *Sturmiopsis inferens* (Order : Diptera) has been successfully used against shoot borer, causing mortalities in the pest for about 25 per cent in Tamil Nadu.

Although several hymenopterous parasitoids found attacking sugarcane borers no large scale use of them has been made possible. However, recently *Isotima javensis* has been successfully mass reared, hence available for use against sugarcane borers in India. It is very good biocontrol agent of *S. nivella. Epiricania melanoleuca* is potent biocontrol agent of pyrilla *Pyrilla purpusilla* (Hemiptera : Fulgoridae). *E. melanoleuca* is Lepidopterous ectoparasite of pyrilla. The parasitoid is difficult to mass rear therefore, it is collected from natural habitat (breeding place) and released into target area for pyrilla control. Due to the efforts taken by Ministry of Agriculture and Co-operative of India, the programme pyrilla control was extended to Maharashtra, Karnataka and Kerala. It is believed that the above ectoparasite is well established in the States of Maharashtra, Karnataka, Andhra Pradesh, Tamil Nadu, Kerala etc. Sugarcane agroecosystem provides good shelter, food and mate for coccineillid beetles. Hence, lady bird beetles like *Chilocorus nigritus* and *Pharoscymnus horni* have been extensively tried for the control of scale insects on sugarcane in Andhra Pradesh and Karnataka.

Biological Control of Sugarcane Wooly Aphid

Sugarcane wooly aphis *Ceratovacuna lanigera* is very bad pest of sugarcane in Kolhapur and Western Maharashtra. *C. lanigera* is reported from oriental and pacific regions of the world including south east Asia, India, Nepal, Bangladesh and Indonesia. In 1900 it was reported very first from Indonesia (Saxena, 1967). In 1974

Ghosh reported it from eastern parts of India including TN and UP. As a sugarcane pest Tripathi (1995) reported this species for the first time as major pest from Asam and Nagaland. Recently, severe out break of *C. lanigera* has been reported from Maharashtra during July 2002 on sugar cane varieties CO-86032 and COC 671 (Patil, 2002, Patil *et al.*, 2004). According to Patil *et al.* (2004) the pest has attained a serious status in several States of India including Maharashtra, Karnataka, Andhra Pradesh, Tamil Nadu, Kerala, U.P., Gujarat etc. In Maharashtra it is till trouble some in Western parts and Marathwada region (Sathe *et al.*, 2009). This pest is difficult to control with pesticides. Acephate 0.1 per cent, Metasystox 0.037 per cent, Diamethoate 0.045 per cent are in use in Maharashtra. However, expected control of the pest is not achieved so far upto date. As natural sources are best and powerful means for control strategies of pests, *C. lanigera* has been surved with its natural enemies from Maharashtra (Table 3.1).

Biological Control Strategies for *C. lanigera*

1. Dipha (*conobathra*) *aphidivora* M. (Figure 3.2) (Lepidoptera : pyralidae) is potential biocontrol agent of *C. lanigera*. Caterpillars of this species feed voraciously on nymphs and adults of *C. lanigera*. Therefore, it is mass reared under shades in field condition and released against *C. lanigera* with 1000/ ha.

2. *Micromus* sp. (Neuroptera : Hemerobidae) is predatory insect. It feed on various stages of sugarcane wooly aphid. The larvae of which look like crocodile are voracious feeders on nymphal stages of the pest. A single larva can consume about 300-400 wooly aphids within 7-9 days. Wooly aphid extend its population on entire leaf of the cane and *Micromus* sp. finish it within 3-4 days if released with 25 to 30 individuals. However, the recommended dose is 2500 individuals per hectare.

3. Syrphid fly (Diptera : Syrphidae) is good biocontrol agent of *C. lanigera*. Syrphid maggots (Figure 3.1) feed on aphid population. The maggots are arayish, tapering anteriorly and broader posteriorly. They catch aphids with mouth and predate upon them. For good control of *C. lanigera*, larvae or cocoons of this predator are released in sugarcane field at the rate of 1000/hectare.

4. Lace wing *Chrysoperla carnea* (Neuroptera : Chrysopidae) is also potential predator of sugarcane wooly aphids. Both larvae and adults of this species feed on sugarcane wooly aphid. The larvae are dirty coloured in young stages and freshwhite in older stages of larval life. The adults are greenish with lace like wings and hence the name lace wing. The adult can survive for 50 days and consume very large amount of aphids. This predator is released on the crop at the rate of 2500 individuals per hectare.

5. Lady bird bettle *Coccinella* sp. (Coleopetera : Coccinellidae) predate, on aphids. Both adults and larvae feed on aphid population. Adults are semicircular, orange red coloured with black spots on elytra. The larvae are with strong madibles, large thorax and tapering abdomen. Strumae are scattered on the body.

Figure 3.1: *C. lanigera* with Syrphid Larvae.

Figure 3.2: *C. lanigera* with Dipha (*Conobathra*) *aphidivora*.

Sugarcane is one of the persistant ecosystem in agriculture. Hence, biocontrol programmes have more chances of success in sugarcane ecosystem. Secondly, the biocontrol method is pollution free, ecofriendly, less expensive and solving permanent solution for the pests.

References

Aggarwal, R.A., 1980. Integrated control of insect pest complex in sugarcane. *Indian Sug.* 29: 649–123.

Patil, A.S., 2002. A new and first recorded sugarcane pest in Maharashtra, white wooly sugarcane aphid *Ceratovacuna lanigera* Zehnter. *Proc. State Level Workshop on a New Sugarcane Pest, White Wooly Sugarcane Aphid Control*, VSI, Pune, pp. 1–13.

Patil, A.S., Shide, V.D., Magar, S.B., Yadhav, R.G. and Nerkar, Y.S., 2002. Sugarcane wooly aphid *Ceratovacuna lanigera* Zehnter its history and control measures. *Co-operative Sugar*, 34: 1–4.

Sathe, T.V., 2004. *Vermiculture and Organic Farming*. Daya Publishing House, New Delhi, pp.1–120.

Sathe T.V. and Bhoje, P.M. *Biological Control of Insect Pests*. Daya Publishing House, New Delhi, pp. 1–122.

Sathe, T.V. and Bhosale, Y.A., 2001. *Insect Pest Predators*. Daya Publishing House, New Delhi, pp. 1–167.

Sathe, T.V. and Jadhav, B.V. *Indian Pest Aphids*. Daya Publishing House, New Delhi, pp.1–214.

Sathe, T.V., K.P.Shinde, A.L.Shaikh and D.K.Raut 2009. *Sugarcane pests and diseases*. Mangl. Publ. Delhi. pp.1–179.

Sathe, T.V., Inamdar, Shakera and Dawale, R.K., 2003. *Indian Pest Parasitoids*. Daya Publishing House, New Delhi, pp. 1–197.

Saxena, A.P., 1967. Biology of *Campyloneurus mutator* (Fab.) (Hymenoptera : Braconidae). *Tech. Bull. CIBC*, 9: 61–72.

Tripathi, G.M., 1995. Record of parasite and predator complex of sugarcane wooly aphid.

Chapter 4

BIOLOGICAL CONTROL OF INSECT PESTS IN MULBERRY ECOSYSTEM

T.V. Sathe*

*Department of Zoology, Shivaji University,
Kolhapur – 416 004, Maharashtra, India*

ABSTRACT

Mulberry pest management is very sensitive issue in sericulture since mulberry leaves is the food of mulberry silkworm *Bombyx mori* L. It should reach to silkworm safety and without contamination of pesticides. Secondly, most of the insect pests of mulberry have developed resistance to the conventional insecticides and their specificity has also restricted the use on mulberry ecosystem. Therefore, ecofriendly control of mulberry pests is the need of the day. Biological insect pest control is a good alternative to insecticidal pest control in mulberry ecosystem. Hence, fisibility and potiential of biocontrol technique against mulberry pests have been discussed in the text.

Keywords: Biocontrol, Mulberry pests, Sericulture.

Introduction

Sericulture in India and at global scenario is emerging very fastly as a good source of currency. As compare to other countries in the world the trend of sericultural

* E-mail: profdrtvsathe@rediffmail.com

development in India clearly depicts a quantum jump in mulberry silk production during last two decades. However, expected yield of the crop is not achived sofar because of pest damage to silkworm and mulberry plant. Sriculture is based on moriculture. Moriculture is cultivation of mulberry for rearing of mulberry silkworkm *Bombyx mori* L. Mulberry is cultivated in more than 30 tropical and subtropical countries. The More common species found in India are *Morus alba* and *M. indica*. In India several varieties are practiced. However, expected yield of crop is not achieved for far because of the insect damage to mulberry plant. So far over 300 species of insect pests have been reported on mulberry from various parts of the world. From India about 100 insect pests have been reported on various varieties of mulberry. They cause the damage to mulberry by defoliating, sucking cell sap, boring the stem and roots and even contaminating crop by various ways.

Since the advent of chemical insecticides, the modern farming methods have become heavily dependent on chemical insecticides which not only increase the cost of production but also results in soil degradation and productivity staganation. Therefore, it is very essential to developed an alternative for chemical insecticides.

The break through in sericultural development was mainly due to the introduction of many high yielding varieties and by birds and adoption of improved crop production techniques. There has been substantial increase in the production of leaf yield and cocoon quality but hybrid and high yielding varieties led to pest problems.

All chemicals affect the environment. Out of which insecticides, Weedicides, Fungicides. Acaricides, Nematicides and Molluscides always increase pest problems and have tremendous pressure on mulberry ecosystem which leads very serious problems such as Air, Water, Soil pollution, health hazards, killing of silkworms and other beneficial insects (such as parasitoids, predators, honey bees), pest resistance, pest resurgence, secondary pest out break, inturrption to food web and ecocycles etc. The Govt. of India has already received 49 pesticides which have been found to cause hazardous to the human health, animals and environment. Out of which 19 pesticides are banned and 11 are restricted to their use. The important pesticides banned by the Govt. of India refers to aldrin, calcium cyanide, copper acetoar senate, chlordane, diabromochloropropene, endrin, ethyl mercury chloride, ethyl parathob, heptachlor menozona, methyl 24 per cent formulation, nicotine sulphate, paraquate dimethyl, sulphate, penta chlorio-nitrogen zene, penta chlorophenol, phenyl mercury acetate, sodium methane arsenate, tetradifon toxaphene, phosphamidon (85 per cent SL), methomyl 12.5 per cent, aldicarb, chlorobenzilate, dieldrin, ethylene dibromide, maleic hydrazide, tricholoro acetic acid, etc. Another 15 pesticides which have been banned resticted in some developed counties but are still being used in India have been subjected for review. The number of registered products has been decresed from 1700 based on more than 300 active ingredients to about 900 formulated products derived from about 200 active ingradients.In USA EPA (The environment protection agency) reviewed the pesticidal registration for banning harmful pesticides to sale.

For tackling pest problem, farmers are largerly dependent on the use of pesticides in various agroecosystems. As a consequence of the indescriminate use of chemical pesticides many sericultural problems have been evolved. The unregulated and excessive use of pesticides has become a major bottle neak in our fight against insect pests. Most of the important species have by now developed resistance to at least one of the insecticides. More than 500 species of insects and mites from agroecosystem have developed resistance to one or another insecticide.In almost all the food material of silk warms and including food grains, vegetables, fruits, meat, fishes, eggs, milk and milk products and even human milk insecticudes have been detected. This clearly indicates that there is need to find out alternative for pesticidal use. Contamination and ventilation are very import prerequisites of sericulture on which the entires success of sericulture business is based. Silkworms are very sensitive to environment and pesticides. Hence, pesticide avoidance has given top priorirty in sericulture. Infact sericulture crops need more ecofriendly climate. Therefore, biological pest control in sericulture has tremendous importance. Biological control is action of parasitoids and predators against target pest for suppression of population below the level of economic damage by keeping environment harmonized.

In past Shimizu (1932), Rohwar (1934), Watanabe (1939), Kuwana *et al.* (1940), Kerrich (1960), Tuhan and Pawar (1983), Qian *et al.* (1984), Samson and Ramadevi (1985), Raynand and Crouzet (1985), Sathe (1987) Kumar *et al.* (1989, 1991), Sathe (1991, 1998, 2004), Singh and Jalali (1991, 1994), Singh *et al.* (1994) Singh and Thangavelu (1995), Siddegowada *et al.* (1995), Geeta Bai *et al.* (1997), Veeranna (1998), Rajadurai *et al.* (1999), Katiyar *et al.* (1999), Sathe and Bhoje (2000), Sathe and Jadhv (2001), Singh *et al.* (2001), Rajadurai *et al.* ((2002), Ram Kishore *et al.* (2002), Jadhv *et al.* (2002), Singh and Saratchandra (2003), Sathe *et al.* (2003), Singh *et al.* (2005) etc attempted the work related to biological control of sericultural pests.

Materials and Methods

In biological control of insect pests four methods are adopted

1. Collection and storage or handling the pests to kill.
2. Collection of parasitoids and predators from a natural habitat and releasing them into target area.
3. Mass rearing and periodic colonization
4. Importation of parasitoids and predators, their mass rearing and colonization in target area.

Parasitoids and predators can be reared as per the techniques given by Sathe (2002,2004) Solayappan (1980), Rao and Devid (1958), Singh *et al.* (2005), Rajadurai *et al.* (2002), etc. Pathogenic biocontrol agents are studied under separate head *i.e.* microbial control. Therefore, parasitoids and predators are the only biocontrol agents left in true biological control of insect pests.

Results and Discussion

Parasitoids

Parasitoids are entomophagus insects and are different from true parasites since, the hosts of parasitoids are restricted to class insecta only. Secondly, the parasitoid always kill its host during or until completion of its life cycle. Parasitoids are scattered in several orders of class insecta. Parasitoids are largely visualized from the orders Hymenoptera, Diptera, lepidoptera and Hemiptera, etc. According to Kerrich (1960) the existing number of parasitoids from insects is 15 per cent of the total population of insects.Order Hymenoptera ranks first in diversity of parasitoids. The family Ichneumonidae alone contain 60,000 species and the family Braconidae 40,000.From Diptera the family Tachinidae contains 1500 species as exclusively parasitic forms.Several species of parasitoids live in mulberry ecosystem. They should be identified and used properly as natural or biological control strategy.

Predators

Predators are generally larger than their preys and they kill the prey by hunting and suppress the pest populations. They consume several preys (pests) in their life. Insect predators are widely scattered in several groups of animals such as Mammals, birds, reptiles, amphibians and pisces from vertebrates and coelentrates, Nematods, helminthes, protozoans, spirders and insects etc. from invertabrates. Birds have very significant role in biological pest control. Wood peckers wood warbles, crows Indian mynha and certain migratory birds largely subsists on mulberry insect pests.From mammalia shrews have tremondous potential in utilization in biological control programme of forest and mulberry pests.Since, they feed on the pupae of several insects of mulberry.

Similarly, jackals, wild dogs, domestic dogs, mongoose cats, etc killed inect diet specially grasshoppers termites, pupae of beeteles and lepidopterous pests.

Indian myhan bird, *Acridotheres tristis* picked cutworms (*Spodoptera litura*), grasshopper and termites from the mulberry ecosystems. The toads *Bufo* spp. are very potential biocontrol agents of mulberry pests. They feed on grasshoppers, crickets and terimites of mulberry crop. In Australia *Bufo marinus* is mass reared and used in biological control of some insect pests. In India, there is need to exploite *B. marinus* on large scale in biological control of mulberry pests. (Sathe, 1998, Sathe and Jadhav 2001, Sathe and Bhoje, 2000).

There are several pests which has successfully controlled by biological means in India by using either parasitoids or predators.

Bihar hairy caterpillar *Spilosoma obloqua* walk (Lepidoptera : Aretidae) (Figure 4.1) is serious pest of mulberry. It occur frequently on mulberry from August to February and and cause serves damage to mulberry by feeding upon leaves. They eat away the chlorophil bearing tissues having exposed veins that affect the photosynthesis of the plant and quality of leaf. The affected leaves dry soon (Figure 4.2) and are easily shed. The caterpillars feed gregariously in first few instars. Attempt have been made to control *S.obliqua* caterpillars in mulberry ecosystem by releasing a braconid larval parasitoid. *Meteorus dichomeridis* (Figures 4.3 and 4.4). It may be released at 100,000/ha/week/twice in a year on mulberry garden. *Meteorus dichomeridis* (Braconidae : Hymenoptera) is mass reared by Sathe 2002 and CSR and TI; Mysore. The above pest is also attacked by *Meteorus spilosomae* (L.), (Braconidae : Hymenoptera) *Apanteles obliquae* (L.), (Braconidae : Hymenoptera), *Glyptapanteles obliqua*, a lerval parasitoid

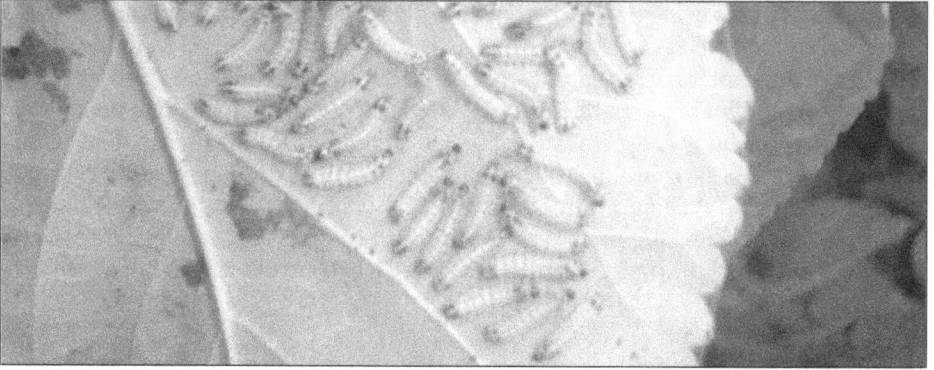

Figure 4.1: *S. obliqua* Larvae.

Figure 4.2: Mulberry Damage by *S. obliqua.*

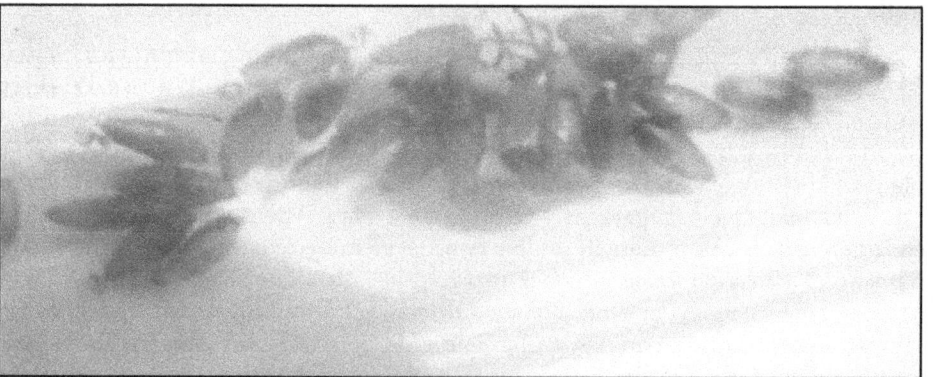

Figure 4.3: *M. dichomeridis* Cocoons

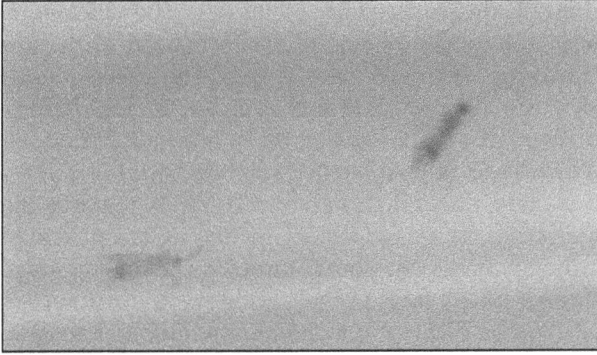

Figure 4.4: *M. dichomeridis* **Adults.**

(L.), (Braconidae : Hymenotera), *A. ceraonoti* (Larval parasite), *Cotesia flavipes* (L.), *Trichogramma minutum* egg parasitoid (Tricogrammatidae : Hymenoptera) and a tachinid fly *Drino* sp. (larval parasitoid) (Tachinidae : Diptera).

Cutworms *Spodoptera litura* (Noctuidae:Lepidoptera) is occurring on mulberry mostly from August to February, caterpillars cause the damage to shoots of young plants by cutting them. The cut portion of the shoot dries up and finally falls down. The caterpillars also feed on the leaves.The Pest can be controlled by introducing *Trichogramma evanescens minutum*, 1,50,000 per hactre as per the need (Sathe, 2004b). Other biocontrol agent refers to Braconids belongs to family Braconidae of order Hymenoptera, *Microplitis prodeniae* (L.), *Apanteles prodeniae* (L.), *A. asawari* (L.) and *Cotesia colemani* (L.) (Sathe, 1987), Sathe (1991). The caterpillars are also parasitized by Ichneumonid flies *Campoletis chlorideae* (L), *Diadegma argenteopilosa* (L.) *Netelia ferruginea* (L.) and *Trichospilus pupivora* (L.) out of which *C. chlorideae* (Figure 1.9) cause about 60 per cent mortality in the caterpillars. *D. argenteopilosa* is also quiet potential biocontrol agent of the above pest. *Netelia* and *Trichospilus* attack older instars of the pest and acts as very good complementory parasitoids for causing mortality in larval stages of the pest. An Eulophid *Tetrastichus ayyari* also parasistizes the larval forms. The cutworm is predated by sting bug *Canthe conva furcellata* (L.), (Hemiptera), *Rhinocoris fuscipes* (L.) (Ruduviidae: Hemiptera) carabid beetle *Parena nigrolineata* (L.), (Scarabaeidae : Coeleoptera) and a Bacteria - *Servatia macrescens* (L.) out of above biocontrol agents sting bug was dominant.

Moringa caterpillar mostly occur on the mulberry from August to March and cause the damage to mulberry leaves by feeding upon them. They remove the solid green content of the leaves as a result leaves start drying, turn yellow and drop down. In seviours infestation entire leaves are defoliated by moringa caterpillars. These caterpillars can be controlled upto certain extent by encouraging *Dolichogenidea parijatki* (Braconidae : hymenoptera) (Sathe *et al.*, 2003) against the caterpillars. The parasitoid attack 3rd and 4th instars of the caterpillar.

Mealybug *Maconellicoccus hirsutus* (Pseudococcidae: Hemiptera), occur mostly in summer season on mulberry garden Mealy bug is considered as the most serious pest of mulberry as it is responsible for "Tukra disease" in mulberry. The nymphs and females are the only destructive form of this pest. They suck the cell sap from the tender leaves and buds which results in malformation of the apical shoot. The malformed head became very hard and compact. The affected leaves get wrinkled and became dark green. As a result of such damage the leaf yield is tremendously reduced and nutritive value of the leaves is badly affected. The above notorious pest can be controlled by releasing the Australian ladybird beetel *Cryptolaemus montrouzieri* @250 adults per hectare. Australian ladybird beetle is rearable under laboratory condition on mealybugs. In the laboratory mealybug are provided cucumber for feeding purpose, the cucumber can store natural food *i.e.* liquid juice for longer period hence it is easy to rear mealy bugs in the laboratory and further lady bird beetles. Mealy bugs are also predated by praying mantis in mulberry ecosystems.

Table 4.1: Natural Enemies of *Diaphania pulverulentalis*.

Host Stage	Natural Enemy	Order and Family	Per cent Incidence*
Egg	I. Parasitoid		
	Trichogramma chilonis	Hymenoptera: Trichogrammatidae	21.18±6.18
	II. Predator		
	Menochilus sexmaculatus	Coleoplera: Coccinellidae	12.09±7.27
Larva	I. Parasitoids		
	Apanteles taragamae	Hymenophera : Braconidae	34.41±14.25
	A. machaeralis	Hymenophera : Braconidae	29.04±13.19
	Apanteles sp.	Hymenophera : Braconidae	22.18±12.34
	Bracon hebetor	Hymenophera : Braconidae	18.17±9.15
	Goniozus indicus	Hymenoptera : Bethylidae	23.10±8.45
	Phanerotoma hendecasisella	Hymenophera : Braconidae	16.32±7.36
	II. Predators		
	Calasoma sp.	Coleoptera : Carabidae	41.36±15.07
	Eocanthecona furcellata	Heteropera : Pentatomidae	14.89±9.28
	Rihirbus trochantrichus	Heteroplera : Reduviidae	9.44±3.56
	Solenopsis sp.	Hymenoptera : Formicidae	6.12±2.341
	Oxyopus sp.	Arachinidia : Lycosidae	10.06±4.31
Pupa	I. Parasitoid		
	Tetrastichus howardii	Hymenoptera : Eulophidae	27.91±11.36
	II. Predator		
	Unidentified earwigs	Dermaptera : Forficulidae	10.12±7.66

* Mean ± SD values.

Mulberry leaf roller (Pyralidae : Lepidoptera) occur on the mulberry in mansoon season and cause the damage to mulberry leaves by rolling and feeding upon them. As a result leaf quality adversely affected. It is one of the key pest of mulberry in southern sericultural states *viz.* Karanataka, Andhra Pradesh and Tamil Nadu causes a leaf yield loss of 12.8 per cent with an average incidence of 21.77 per cent (Siddegowda *et al.*, 1995, Rajaduria *et al.*, 1999, Jadhav *et al.*, 2002). The pest can be controlled by introducing *Trichogramma chelonis* 1,50,000 per hactare. Caterpillars are only destructive forms. Caterpillars get attacked by more than 12 species of insect parasitids and predators. Their pupae also paraistized and predated by an *Eulophid* and a *forficulid* respectively The natural enemies recorded by Rajadurai *et al.* (2002) on *D. pulverulentalis* are reported in Tables 4.1 and 4.2.

Dragonflies have great potential in pest suppression in mulberry ecosystem. Sathe *et al.* (2006) reported 12 species of dragonflies feeding on noctuid moths pyralid moths, jassids, aphids, grasshoppers, beetles, thrips, etc. Out of 12 species reported from mulberry ecosystem *Crocothemis servilia* (Drury) and *Pantala flaviscens* (Pabricius) were most dominant in Maharashtra specially western Maharashtra. The incidence of parasitoids and predators was ranged between 16.32-34.41 per cent and 6.12 – 41.36 per cent respectively. Among the parasitids and predators. *Apanteles taragamae* and *Calasoma* sp. showed the highest incidence respectively. As regards the parasitic and predatory potential, the laboratory studies have revealed higher parasitization portential by *T. chilonis* (77.83 per cent) and *A. taragamae* (71.18 per cent) compaired to other parasitoids. Among the predators, *Calasoma* sp. has recored the highest (66.67 per cent) (Table 4.2).

Table 4.2: Parasitization and Predatory Potential of Natural Enemies of Leaf Roller.

Natural Enemy	Name of the Parasifoids/ Predators	Stage of Pest	Per cent parasitization/ predation*
Prasitioid	Trichogramma chilonis	Egg	77.83±6.18
	Apanteles taragamae	Larva	71.18±4.37
	A. machaeralis	Larva	65.80±4.13
	Bracon hebetor	Larva	52.24±7.19
	Goniozus indicus	Larva	42.90±4.36
	Tetrastichus howardii	Pupa	58.87±7.96
Predator	Calasoma	Larva	66.67±9.10
	Eocanthecona furcellata	Larva	57.63±10.32
	Menochilus sexmaculatus	Egg	40.18±4.36

* Mean ± SD values.

According to Rajadurai *et al.* (2002) a total of 15 species of parasitoid and predatory insects belonging to four orders and 11 families were found to attack three different stages (egg, larva and adult) of leaf roller. Rajaduria *et al.* (1999) reported for the first time in India about the biocontrol programme in IPM against this pest which includes release of *Tetra stichus howardii* – indigeous gregarious hymenopteran pupal parasitoid and *Trichogramma chilonis* an egg parasitoid.Geetha Bai *et al.* (1997) mentioned that incidence of *Phanerotoma noyesi* was heavy (5-39 per cent) whereas *Apanteles* sp. was low (0-2 per cent). But in contrast in the present observation, *Apanteles* sp. exhibited high incidence rate (34.41 per cent) than *Phanerotoma* sp. (16.32 per cent).

Kuwana *et al.* (1940) have studied the biology of *Macrocentrus philippinesis;* Watanabe (1939) has presented a list of 16 hymenopteran parasitoids and Veeranna (1998) have reported the natural enemies of leaf roller in China *viz.*, *Trichogramma evanescens*, *T. matsumura* the egg parasitoids and *Apanteles heterusiae* and *Macrocentrus philipinensis* as larval parasitoids.

Above reported 15 natural enemies on leaf roller, which can be exploited as a biocontrol agents of the pest in the mulberry crop system. The availability of natural enemies of _D. pulverulentalis_ provides immense opportunity to pest control specialists. Further, the potential natural enemies can serve as supportive and/or suitable alternatives to the available management strategies against the leaf roller.

Biocontrol approaches of pest are not only useful against mulberry pest but also for the non mulberry pests. Ram Kishore _et al._ (2002) studied the role of _Trichogramma chilonis_ Ishii in suppression of field population _Notolophus antiquae_. The vapourer tussock moth caterpillar _Notolophus antique_ (Lymantriidiae : Lepidoptera) is a sporadic pest of primary tasar food plants _Terminalia arjuna_ Bedd and _T. tomentosa_ and during Mansoon period in the tasar culture area. An egg parasitoid _Trichogramma chilonis_ was found a potential biocontrol agent against _N.antique_ in tasar culture. The parasitization of 35.55 per cent eggs of _N.antigua_ was observed when _T. chilonis_ was released in tasar food plant garden @ 2,00,000 adults per hactare at an interval of 3 days. It destroyed an appreciable population of _N. antique_ eggs. Significantly higher parasitization was recorded in the direction of wind compared to the opposite direction.

Ram Kishore _et al._ (2002) used _T. chilonis_ against _N. antique_ eggs in the tasar food plants ecosystem, for the first time. In sunflower _Spilosoma obliqua_ was reported to be the host for _T. chilonis_ but no information is available on field efficacy of this parasitoid (Singh and Jalali, 1994) _T. chilonis_ has been exploited as a biocontrol agent and its field efficacy has been tested against several agricultural pests. The rate of parasitization in other pests by _T. chilonis_ ranged from 12.6 to 90.0 per cent. Higher parasitization in majority of these cases by _T. chilonis_ as compared to _N. antique_, may be due to uniform distribution of host eggs in the respective field crops where the host eggs become easily available to the parasitoid. By contrast in the case of _N. antique_ the eggs are laid in batches on the under surface of the host plant leaves in isolated locations, thereby rendering the host eggs less accessible to the parasitoid. They further reported that _T. chilonis_ was found to parasitise _N. antique_ caterpillar eggs upto 8 m radius though it reached up to 15 m radius when assisted by wind. Singh and Jalali (1991) and Katiyar _et al._ (1999) have reported that _T. chilonis_ parasitizes eggs of _S. obliqua_ up to a radius of 10 m. Host searching up to a distance of 100 m by _T. chilonis_ from release points has been reported by Tuhan and Pawar (1983). Other species _viz., Trichogramma maidis_ and _T. ostriniae_ were able to parasitise _S. obliqua_ eggs upto 15 m (Raynand and Crouzet, 1985) and 20 m (Qian et al, 1984), respectively.

The work of Ram Khoshore _et al._ (2002) demonstrate that parasitization of _N. antique_ eggs by _T. Chilonis_ is higher with the wind direction than in the opposite direction. This is because of the fact that the parasitoid being minute in size, is not strong enough to more against the wind current over long distance especially if wind velocity is high. On the other hand if wind velocity is high, the parasitioid almost gets drifted away in the wind direction when released in the field, thereby reaching longer distance for parasitizing the host eggs. It is, therefore obvious that in case of egg parasitoids, the role of wind current is most important especially if the parasitioid has to move upwind to parasitise the host eggs. Similar observations were reported by Katiyar _et al._ (1999) in case of _S.oblique._

As regards to biocontrol approaches of silkworm pests, very little attention is paid except the work on *N. thymus*, a biocontrol agent of uzi flies *Exorista bombycis Leuis* and *Blolepharipa zebina*. The above uzi flies are the larval endoparasitoids of mulberry silkworm, *Bombyx mori* (L.), Tasar Silkworm *Antheraea mylitta* Drury, Eri silkworm *Samia cynthia ricini* Bosidual and muga silk worm *Antheraea assama* westwood.

Singh *et al.* (2005) studied augmentation of natural enemies to control uzi fly in serculturre. They reported that several parasitoids of uzi flies namely.*Nesolynx thymus, Dirhinus* spp., *Trichopria* spp. and *Trichomalopsis* spp. have been screened and found effective against maggot and pupal stages of uzi flies. The uzi fly populations were brought down considerably (60 per cent) through release of the parasitoids. Under Integrated Pest Management (IPM) Programme of uzi flies, biological control is considered one of the important components. Biological potential, reproductive strategy, host searching ability and sex ratio of the parasitoids have been reported. Biological suppression methods involving augmentation, conservation and utilization of natural enemies are the most important aspects and the pest populations are maintained at lower levels by the action of parasitoids in their natural habitats. Augmentation involves actions to increase the populations of beneficial effects of parasitoids. Augmentation through inoculative or inundative releases of parasitoid is the most direct way of increasing the numbers of these beneficial organisms. However, our limited mass rearing capability is a serious impediment to the effective implementation of augmentation programme. Suitable strategies have been evolved for mass rearing of the parasitoids. Efficient and economical rearing systems are the key requirements to wide spread use of this approach in biological control. Cost efficient and effective utilization of augmentation techniques are possible and have been implemented successfully to contain the fly pest populations.

Both predators and parasitoids are the animals that feed on the other animals. The predator consumes several individuals (prey) during its development whereas the parasitoid completes its development on a single host (Singh *et al.*, 2001). parasitoids that have been screened and their parasitization potential tested against uzi fly pupae (*E.bombycis*) include *Nesolynx thymus, Trichopria* sp., *Exoristobia philippinensis, Dirhinus* sp., *Brachymeria lugubris, Spalangia emlius, Pachycrepoideus veemnnai* and *Spilomicrus karnatakensis* (Samson and Ramadevi, 1985; Kumar *et al.*, 1989; 1991). Similarly, *Trichomalopsis apanteloctena, Pediobius* sp., *Brachymeria lasus* and *Theronia maskaliva* have been reported to parasitize tasar uzi fly pupae (Singh etal., 1995; Singh and Thangavelu. 1996). Though almost all of them are found effective in minimizing uzi fly population, their parasitization potential differs considerably.

Among the several natural enemies (parasitoids) reported to attack the uzi fly, *E. bombycis*, only a few of them *viz., N. thymus, E. philippinensis, Trichopria* sp. and *Dirhius* sp. have been studied in some detail. *N.* thymus has been chosen as a biocontrol agent as it possesses most of the desirable attributes of a biocontrol agent (Kumar *et al.*, 1996). However, before exploitation of *N. thymus* on a commercial scale, certain approaches to applied, biological control like conservation and augmentation of the parasitoids have to be made to render the biocontrol of *E. bombycis* more meaningful apart from working out the possibilities of importation of the parasitoid (s) from the native land of the pest.

The following are the attributes that are considered desirable while chosing a natural enemy/biocontrol agent (parasitoid and/or predator) in the pest suppression programme. a) Efficient host/prey searching ability, b) superior parasitisation/ predation ability, c) better synchronisation of life cycle with that of host/prey, d) shorter life cycle, e) amenability to laboratory production, f) better temperature tolerance, g) high degree of host/prey specificity, h) good adoption to survive under field conditons, etc. Although several potential parasitoids of uzi fly are existing in the nature, exploitation of these agents has not been done due to lack of sufficient information on the behavioral response of the parasitoids. Several hymenopterous parasitoids are available in nature, which can be utilized to control the uzi fly populations. These are considered as potential weapons and, therefore, can be exploited as biological control tools (Singh and Saratchandra, 2003).

References

Geetha Bai, M., Marimadaiah, B., Narayanswamy, K.C. and Rajagopal, D., 1997. An Outbreak of leaf roller pest, *Diaphania* (=Margaronia) *pulverulentalis* (Hampson) on mulberry in Karnataka. *Geobios New Reports*, 19: 73–79.

Jadhav, A.D., Bhoje, P.M. and Sathe, T.V., 2002. Reproductive potential of *Apanteles taragamae* Viereck in relation to age of Mulberry leaf roller, *Diaphania pulverulentalis* (Havipson). *6th Cong. Int. Seri. Com. Proc.*, 1: 79–83.

Katiyar, R.L., Yogananda, M.C., Manjunath, D., Sen, A.K., Ahsan, M.M. and Datta, R.K., 1999 Role of *Trichogramma chilonis* Ishis in the suppression of field population of the Bihar hairy caterpillar *Spilosoma obliqu* Walker. *Indian J Seric.*, 38: 40–43.

Kerrich, G.J., 1960. The state of our knowledge of the systematics of the *Hymenoptera parasitica*. *Trans. Soc. Brit. Entomol.*, 14: 1– 18.

Kumar, A., Kumar, P., Singh, B.D. and Sengupta, K., 1991. Parasites of uzi fly *Exorista sorbillans* Wiedemann (Diptera : Tachinidae) VI. Record of a new *Dirhinus* sp. on uzi fly and notes on its bilogy. *Sericologia*, 31: 251–252.

Kumar, P., Kumar, A., Noamani, M.K.R. and Sengupta, K., 1989. Parasite of uzi fly *Exorista sorbillans* Wiedemann (Diptera : Tachinidae) a new record. *Curr. Sci.*, 58: 821–822.

Kuwana, Z., Ishi, G. and Kurosawa, T., 1940. Studies on the Hymenopterous parasites of *Margaronia pyloalis* Walk. (Lepidoptera) Parasite *Macrocentrus philippinensis* Ash. (Braconidae). *Rep. Seric. Exp. Sta. Tokyo*, 10: 1–26.

Qian, Y.Q., Cao, R.L. and Li, G.Z., 1984. Biology of *Trichogramma ostrinide* and evaluation of its effectiveness in controlling corn borer on spring corn. *Acta Entomol Sin*, 27: 287–293.

Rajadurai, S., Sen, A.K., Manjunath, D. and Dutta, R.K., 2002. Natural enemy fauna of Mulberry leaf roller, *Diaphania pulverulentalis* (flampson) (Lepidotera : Pyralidae) and its potential. *Adv. Indian Ser. Res.*, 13: 220–223.

Rajadurdi, S., Manjunath, D., Katiyar, R.L., Prasad, K.S., Sen, A.K., Shekha, M.A., Ahsan, M.M. and Datta, R.K., 1999. Leaf roller: A serious pest of mulberry. *Indian Silk*, 37(12): 9–11.

Ram Kishore, S.P., Sharama, S.K., Sharan, S.K., Kulshreshtha, Varun and Thangavelu, K., 2002. Role of *Trichogramma chilonis* in the suppression of field populations of *Notolophus antique* Linn. *Adv. Indian. Ser. Res.*, 5: 153–156.

Rao, T.V., 1958. The biological control of coccid pest in south India by the use of the beete, *Crypholaemus nontrouzieri* Muls. *Indian J. Agric. Sci.*, 28: 545–552.

Raynand, B. and Crouzet, B., 1985. Maize control of pyralid with *Trichogramma phytoma*, 366: 17–18.

Rohwer, S.A., 1934. Descriptions of five hymenophteran parasites. *Proc. Ent. Soc. Washingaton*, 36: 43–48.

Samson, M.V. and Ramadevi, O.K., 1985. A new record on the parasites of the uzi fly *Tricholyga bombycis* Beck. *Curr. Sci.*, 54: 11-44.

Sathe T.V., 1987. New reords of natural enemies of *Spodoptera litura* (Fab) in Kolhapur, India. *Curr. Sci.*, 56(20): 1083–1084.

Sathe, T.V., 1991. The biology of *Apanteles asawari* Sathe (Hymenoptera : Braconidae) a larval parasitoid of *Spodoptera litura*. *Oikoassay*, 8(1 and 2): 15–18.

Sathe, T.V., 2004. Biology and behariour of coccinelid beetles. *Indian Insect Predators in Biological Control*, 8: 175–203.

Sathe, T.V. and Jadhav, A.D., 2001. *Sericulture and Pest Management*. DPH, New Delhi, pp. 1–165.

Sathe, T.V. and Bhoje, P.M., 2000. *Biological Control of Intesct Pests*, DPH, New Delhi, pp. 1–121.

Sathe, T.V., Shinde, K.P., Jadhav, B.V. and Chougle, T.N., 2006. Role of dragonflies for suppression of pests in mulberry ecosystem in Kolhapur, India. *Proc. Asia Pacific Cong. Seri. and Biotech*, 6: 159.

Sathe, T.V., Inamdar, S.A. and Dawale, R.K., 2003. *Indian Pest Parasitoids*, DPH, New Delhi, pp. 1–157.

Sathe, T.V., 1998. *Sericultural Crop Protection*. Asawari Pub. Osba., pp. 1–122.

Sathe, T.V., 2002. Intrinsic rates of increase and interspecific relationship between *Meteorus dichomeridis* and *Meteorus spilosomae* the larval parasitoids of mulberry pest *Spilosama obliqua*. *19th Cong. Int. Seri. Comm. Proc.*, 1: 39–46.

Sathe, T.V., 2004b. *Vermiculture and Organic Farming*. DPH, New Delhi, pp. 1–122.

Shimizu, M., 1932. Oviposition of *Epiurus* (Pimpla) *persimilis* Ashm. a parasite of *Margaronia pyloatis* Walk. *Kontyu*, 6: 169.

Siddegowda, D.K., Gupta, V.K., Sen, A.K., Benchamin, A.K., Manjunath, K.V., Prasad, D., Madgum, K.S., S.B. and Datta, R.K., 1995. *Diaphania* species infests mulberry in south India. *Indian Silk*, 34(12): 6.

Singh, R.N. and Thangavelu, K., 1995. First record of *Trichomalopsis apanteloctena* on the uzi fly. *Indian J. Seric.*, 34: 164–165.

Singh, R.N. and Thangavelu, K., 1996. Biological characteristic of *Trichomalopsis apanteloctena* Crawford, a parasitoid of *Blepharipa zebina*. *Indian J. Seric.*, 35: 62–63.

Singh, R.N. and Saraichandra, B., 2003 Biological control strategy of uzi fly in sericulture. *Int J. Indust. Emtomol.*, 6(2): 125–132.

Singh, R.N., Sinha, B.R.R.P. and Sinha, S.S., 1994. Host discrimination between parasitised and unparasitised uzifly pupae by females of *Trichomalopsis apanteloctena* Crawford (Hymenoptera: Pteromalidae). *Trends Life Sci.*, 9: 27–32.

Singh, S.P. and Jalali, S.K., 1991. Trichogrammatid egg parasitoids their production and use. Extension Bulletin, B.C.C.N.C, IPM, Faridabad, India, 1: 15.

Singh, S.P. and Jalali, S.K., 1994. Trichogrammatids. Technical Bulletin, Project Directorate of Biological Control, Indian Council of Agricultural Research, Bangalore, India P.28.

Singh, R.N., Mahesherari, M. and Savatchandra, B., 2005. Augmentation of Natural enemies to control Uzifly in sericulture. *Adv. Trop. Seri.*, 20: 330–333.

Singh. R.N., Krishna Rao, J.V. and Samson, M.V., 2001. Role of parasitoids in pest management in tasar culture. In: *Biocontrol Potential and its Exploitation in Sustainable Agriculture.*

Sotayappan, A.R., 1980. Some new trends in the rearing and field release of *Trichogzamma australicum* Gir, in Madurantakam Sugar factory area Tamil Nadu. *Indian Sug.*, 30: 19–24.

Tuhan, N.C. and Pawar, A.D., 1983. Life history, host suitability and effectiveness of *Trichogramma chilonis* (Ishii) for controlling sugarcane borers in Punjab. *J. Adv. Zool.*, 4: 71–76.

Veeranna, G., 1998. Insect pest of mulberry and their management in China. *Indian Silk*, 36(6): 5–9.

Watanabe, C., 1939. Preliminary notes on Hymenopterous parasites of the mulberry pyralid moth, *Margaronia pyloalis* Walker. *Kontyu*, 5–6: 231–236.

Chapter 5

CARABIDS (COLEOPTERA : CARABIDAE) FOR BIOLOGICAL CONTROL OF INSECT PESTS

T.V. Sathe*

Department of Zoology, Shivaji University,
Kolhapur – 416 004, Maharashtra, India

Introduction

Carabids (Coleoptera : Carabidae) are potential biocontrol agents of various insect pests such as aphids, termites, Lepidopteran pest caterpillars and pupae, coleopteran grubs and pupae, and other many insect pests. They are recorded from various agricultural ecosystems, orchards, plantation, forests etc. Review of literature indicates that many species of carabids are recorded as insect pest predators from India which refer to *Calosoma moderae* Fabricius, *Calleida splendidula* Fabricius, *Parena nigrolineata* Chaudoir, *Casnoidea indica* (Thunberg), *Omphra pilosa* Klug, *Chlaenius pangaeoides* Laferte, etc. In past, Fletcher (1919), Ramchandra Rao (1924), Samal and Mishra (1978), Rajagopal (1984), Rajagopal and Prasad Kumar (1992), Patil and Sathe (2003) etc. worked on biodiversity and biocontrol potential of carabids from India. For the control of forest pest, Gupsy moth *Lymantria dispar* L. the carabid *Calosoma sycophanta* L. was imported on large scale from Europe to North America (Burgess, 1911). For appropriate record of carabids as pest insects and their utility in pest control programmes this chapter will add great relevance.

_____

* E-mail: profdrtvsathe@rediffmail.com

Carabids for Biological Pest Control

The list of carabids for control of various insect pests is given in Table 5.1.

Table 5.1: Carabids for Biological Control of Insect Pests

Sl.No.	Carabid	Insect Pest	Crop
1.	*Calosoma moderae* Fabricius	Aphids, cutworms Caterpillars Termites *Lymantria mathura*	Cotton, paddy, maize Cotton, paddy Horticulture and forest pests Horticulture and forest pests
2.	*Calosoma sycophanta* L.	*Lymantria dispar* L. *Lymantria mathura*	Forest trees Forest trees
3.	*Calosoma moderae* Fabricius var. *indicum* Hope	*Agrotis ypsilon* *Spodoptera litura* Fab. *Mythimna separata* (Walker) *Plusia* sp.	Gram, Cowpea Cowpea Maize Pulses
4.	*Casnoidea indica* (Thunberg)	Brown plant hopper *Nilaparvata lugens* (Stat.)	Paddy
5.	*Calleida splendidula* Fabricious	Termite *Odontotermes* spp. Aphids Caterpillars	Crops Forest, Agriculture and Horticulture Forest, Agriculture and Horticulture Forest, Agriculture and Horticulture
6.	*Chlaenius pangaeoides* Laferte	Aphids *Aphis craccivora*	Cowpea
7.	*Parena nigrolineata* Chaudoir	Coconut Black head caterpillar *Nephantis serinopa* Meyrick *S. litura*	Coconut Tobacco
8.	*Parena laticineta* (Bates)	*N. serinopa*	Tobacco
9.	*Oxylobus dekkanus* Andrews	*Odontotermes horni* (Wasmann) *O. obesus* (Rambur) *Microtermes obesi* Holmgren	Crops : Forest, Horticultural, Agriculture Crops : Forest, Horticultural, Agriculture Crops : Forest, Horticultural, Agriculture
10.	*Omphra pilosa* Klug	Termites *Odontotermes* spp. *Microtermes* sp.	Crops: Forest, Agriculture, Horticulture Forest, Agriculture, Horticulture
11.	*Pheropsophus hilaris* (Dej)	Rhinoceros beetle *Oryctes rhinoceros*	Coconut
12.	*Phlaeodermius nigrolineatus* (Chand.)	*N. serinopa*	Coconut
13.	*Tetragonoderus* spp.	Grasshoppers *Heiroghyphus banian*	Paddy, Maize, Jowar, Wheat

Contd...

Table 5.1–*Contd...*

Sl.No.	Carabid	Insect Pest	Crop
14.	*Dicranoncus* sp.	Rice bug *Lyptocerisa varicornis* (Thunburg)	Paddy
15.	*Chlaenius rayotus* (Bales)	*Helicoverpa armigera* (Hubn.)	Gram, Red gram, Maize, Jowar, Tomato, Cotton
16.	*Cicindela sexpunctata*	Rice bug *L. varicornis*	Paddy
17.	*Cicindela campestris*	Rice bug *L. varicornis*	Paddy
18.	*Anthia sexguttata* Horn. *A. kolhapurensis* Sathe and Patil	Grasshopper, *H. banian* *H. banian*	Paddy Paddy
19.	*Eudema angulatum* Horn.	Cutworms, Wire worms, Jassids, Aphids, Crickets, Caterpillars, Grasshoppers, Cockroaches	Crops:Agricultural, Forest, Horticultural, Household etc.
20.	*Eudema polyphagi*	Polyphagus, as above	As above
21.	*Dicranoncus indicus* Sathe and Patil	Caterpillars, Jassids aphids, grasshoppers etc.	Crops: Agricultural, Forest, Horticultural
22.	*D.* (*Monacanthonyx*) *pocillator* Horn.	As above	As above

Biology of Carabids

Carabids belongs to family Carabidae of order Coleoptera hence they are complete metamorphic forms (Holometabolus). Thus, all carabids show four distinct stages of life cycle *viz.* egg, larva, pupa and adult. Carabids are biocontrol agents and are exclusively predatory insects. Carabid beetles are confused with cicindelid beetles. However, they are separated by having lateral extension of the clypeus in front of the antennae. Carabid beetles vary in the size from small to moderately large, the smallest one is quarter of an inch long and the largest is nearly one inch in length. The colour is varied from brown to black some times with bright patches of yellow and is often strikingly warning. They are with oval body and broader than cicindelids, more flattened with mostly filiform antennae. The eggs of carabids are rounded/oval, whitish. They hatch with in one or two weeks. The grubs are whitish and predaceous. There are 5 to 6 instars in carabid beetles. The larvae or grubs are slender and active creatures with large head and long mandibles and six ocelli. The thorax and abdomen are smooth and tapering with terminal pair of dorsal cerci and oval tube and 3 pairs of thoracic legs. The grubs prefer caterpillars for feeding purpose. The full grown grub pupates in soil and then transformed into adult beetle. In some species like *Lytta* and *Zonabria postulata* there is hyper metamorphosis in larval stage. The larvae of *Z. postulata* feed on eggs of grasshopper and acts as a good biocontrol agent for grasshoppers. The larvae of carabids are the main predaceous individuals which constitute a part of the surface fauna of pests and best found when caterpillars are abundant on the crop. Imago stage is hibernating stage. Adults burying themselves in the soil or may take shelter. The carabids are partly diurnal and partly nocturnal. They may have even defense gland for emitting disagreeable odour.

Biodiversity of Carabids

The family carabidae is one of the largest family of the order Coleoptera (Patil and Sathe, 2003). About 21,000 species are described under this family from different parts of the world. Very interestingly their elytra is fixed and hind wings are atrophied. From India about 2000 species have been reported (Patil and Sathe, 2003). Biodiversity and taxonomy of carabids have been studied by Andrews (1929, 1930, 1935), Horn (1915), Sloane (1923), Moore (1963), Pearson (1988), Saha *et al.* (1992), Patil and Sathe (2003) etc.

1. *Calosoma moderae* Fabricius var. *indicum* Hope

Distribution	:	Maharashtra, Karnataka, Andhra Pradesh, Tamil Nadu, Punjab etc.
Description	:	Adult - (Length x width) 30' x 11' mm
		Color - Black
		Prorthorax - much wider than head
		Elytral surface - wider at base and imbricate.

2. *Casnoidea indica* (Thunburg)

Distribution	:	Karnatkata, Andhra Pradesh, Maharashtra and Tamil Nadu
Description	:	Adult - (Length x width) 7.2' x 2.00' mm
		Color - Brown
		Prorthorax - Narrowed behind
		Elytral surface - Deep brown with two pairs of transverse black bands

3. *Chlaenius pangaeoides* Laferte

Distribution	:	Maharashtra, Karnataka, Andhra Pradesh
Description	:	Adult - (Length x width) 14.8' x 5.3 mm
		Head - Green with coppery reflections
		Elytra - Black with four distinct yellow patches, punctate
		Pronotum - Punctate

4. *Parena nigrolineata* Chaudoir

Distribution	:	T.N., Maharashtra, Karnataka, Andhra Pradesh, Kerala
Description	:	Adult - (Length x width) 8.8' x 3.4 mm
		Colour - Red
		Elytra - Finely truncate, with black longitudinal bands on lateral sides.
		Life cycle - Completed within 40 days.

5. *Omphra pilosa* Klug

Distribution   :   Maharashtra, Karnataka, Tamil Nadu, Andhra Pradesh, Kerala

Description   :   Adult - (Length x width) 20.0' x 7.10 mm

Colour - Uniformly black

Pronotum - Densely punctate

Elytra - Flat with deep striae and small hairs.

6. *Oxylobus dekkanus* Andrews

Distribution   :   Maharashtra, Karnataka, Tamil Nadu, Andhra Pradesh, Punjab and Madhya Pradesh

Description   :   Adult - (Length x width) 18.5 ' x 5.5 mm

Colour - Black with shiny body surface

Mandibles - with 3-4 strong teeth

Elytra - Oval and convex

Protibia - Flattened and outwardly digitate.

Pronotum - comparatively wide in abdomen proportion.

7. *Anthia kolhapurensis* Sathe and Patil

Distribution   :   Maharashtra, Karnataka.

Description   :   Adult - (Length x width) 45 x 15 mm

Colour - Black

Antenna - 20 mm long

Elytra - 25 mm long, 15 mm wide, black with white distinct spots.

Sipho - 11 mm long.

8. *Anthia sexgutatta*

Distribution   :   Maharashtra, Karnataka, Andhra Pradesh, Tamil Nadu, Kerala, Punjab.

Description   :   Adult - (Length x width) 47 x 16 mm

Colour - Black

Elytra - Black with six white distinct spots.

Sipho - shape and size differs than *A. kolhapurensis*.

9. *Eudema angulatum* Horn. (Figure 5.2)

Distribution   :   Maharashtra, Karnataka, Andhra Pradesh, Tamil Nadu, Madhya Pradesh, Uttar Pradesh

Description   :   Adult - (Length x width) 15 x 7 mm

Colour - Black

Antenna - 9 mm long

Elytra - Black with 4 irregular yellow patches, 10 mm long and 3.5 mm wide

Sipho - 1.98 mm long.

10. *Dicranoncus indicus* Sathe and Patil (Figure 5.1)

Distribution   :   Maharashtra, Madhya Pradesh, Andhra Pradesh, Tamil Nadu, Karnataka,

Description   :   Adult - (Length x width) 10 x 5 mm

Colour - Yellowish reddish brown

Antenna - 7 mm long

Elytra - Black brown, 7 mm long, 2.5 mm wide with two yellowish spots.

Spermathica - 0.35 mm long.

11. *Dicranoncus (Monacanthonyx) pocillator* Horn

Distribution   :   Maharashtra, Karnataka, Tamil Nadu, Andhra Pradesh, Madhya Pradesh and Uttar Pradesh

Description   :   Head and thorax similar to *D. indicus*

Mandible - pointedly triangular.

12. *Dioryche colembensis* Nietner

Distribution   :   Andhra Pradesh, Maharashtra, Karnataka,

Description   :   Beetle consumes cowpea aphids in Kolhapur region.

According to Rajagopal and Prasad Kumar (1992) the grubs and adults of *C. moderae* var. *indicum* were quite active during the months July to November and were nocturnal and adults consumed 3-5 caterpillars and 4 to 6 pupae per day. The adult can survive for about 130 days when provided food and die within 30-45 days without food. This is effective biocontrol agent for pests of maize, specially *Spodoptera* sp. and *Plusia* sp. The beetles of *O. dekkanus* start emerging immediately after the onset of monsoon and its population found increased during the months, May to November. The beetles and grubs preferred several species of termites in Western Ghats and in agro ecosystems in Kolhapur region. The adults can survive from 300 days to 425 days with termite food and with control 30-100 days. The beetles consumed 15 to 35 termite workers per day.

The beetles of *C. indica* were most active from June to December. It is potential biocontrol agent of brown plant hopper (*N. lugens*) in paddy fields of Kolhapur district. Being active fliers, the beetles observed throughout the year and get attracted to light because their prey *N. lugens* which was also attracted to light source while maximum

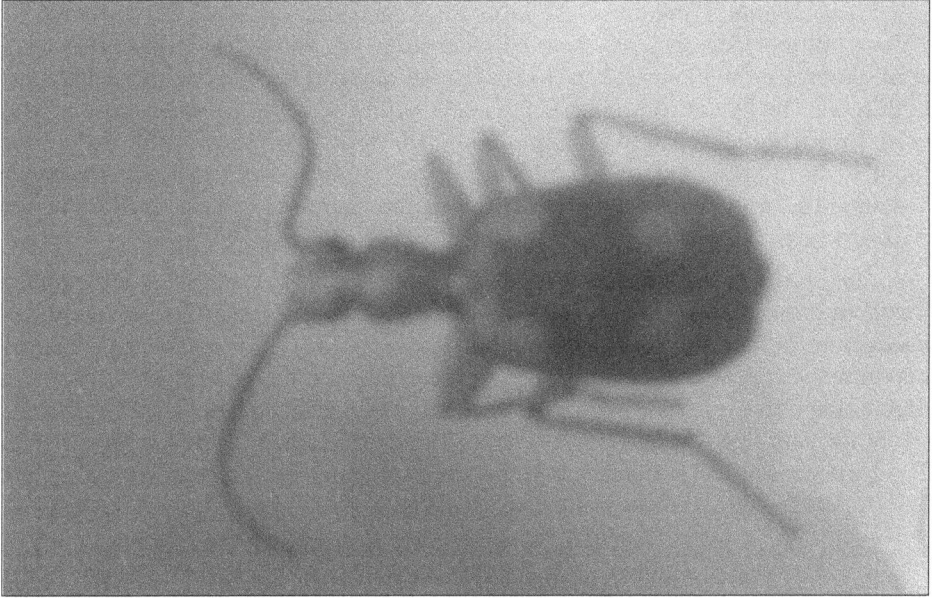

Figure 5.1: *Dicranoncus* sp.

Figure 5.2: *Eudema angulatum.*

population of *O. pilosa* was noted during the months April to October in Kolhapur region. It is abundant in Western Ghats and various agroecosystems of plain Kolhapur region. *O. pilosa* can consume 10-20 worker termites in a single day. Most of the beetles preferred worker caste over others for feeding purpose.

P. nigrolineata is rearable under laboratory conditions (25±1°C, 75 per cent R.H., 12 hr photoperiod) on black headed caterpillar, *O. arenosilla*. The beetles were abundant from the months April to September and can survive for about 450 days with food. The beetle complete its life cycle within 40 days. *C. pangaeoides* beetles occur throughout the year in Kolhapur district. However, maximum population was found in the months May to September. They feed on *Aphis craccivora* Koch. The beetle consumed 4.5 aphids per day (range 2 to 10 aphids per day) and survived for about 30 days (range 20-35 days).

A. sexgutatta is potential predator of grasshoppers, many caterpillars and Aphids. It is abundant in Monsoon season but winter is spend in the holes. This wingless beetle is entirely found on the soil in plain region of India. The beetle can consume maximum 200 grasshoppers per day. The consumption rate is dependant on the predator age and type of grasshopper and age. The eggs of this beetle are large, oval, white and soft. Eggs are laid singly. Oviposition period extends considerably longer. The grubs are whitish and predatory, they feed on caterpillars, aphids, cutworms, etc and pupate in soil. The adults are also predaceous upon above said insect pests.

E. angulatum is widely distributed in Kolhapur region. It consumes about 10-50 grasshoppers 5-30 caterpillars of cutworms and *Spodoptera* per day. Hence, this beetle has good potential in pest control. This beetle can also suppress populations of jassids, crikets, termites, whiteflies and mealy bugs in Kolhapur region.

In paddy ecosystem several predatory beetles are noticed. Predatory carabids are mostly specific and they have high searching ability and pest consumption rate. They can survive for longer period with no food. Soil moisture plays an important role in their population dynamics. Most of carabids are nocturnal. Hence difficult to watch them. However, their survey, identification, conservation and utilization in biological control of insect pests is the need of the day as ecofriendly pest control.

References

Andrews, H.E., 1929. *The Fauna of India including Ceylon and Burma, Coleoptera, Carabidae*, Vol. 1. Taylor and Francis, London.

Andrews, H.E., 1930. *Catalogue of Indian Insects, Part 18, Carabidae*. CPB, Calcutta, Govt. of India.

Andrews, H.E., 1935. *The Fauna of British India including Ceylon and Burma*. Taylor and Francis, London.

Bhat, P.S. and Rajagopal, D., 1988. Carabid fauna of Banglore and their ecology. Part–I. *J. Soil. Biol. Ecol.*, 8: 122–129.

Patil, V.J. and Sathe, T.V., 2003. *Insect Predators and Pest Management*. Daya Publishing House, pp. 1–215.

Pillai, G.B. and Keshava Bhat, S., 1987. Biological and feeding potentiality of *Parena nigrolineata* Chaudoir (Coleoptera : Carabidae), a predator of the coconut caterpillar, *Opisina arenosilla* Walk. *Indian Coconut J.*, 17: 1–5.

Rao, G.A., Rao, P.R.M., Laxminarayana, K. and Rao, P.S., 1978. Preliminary observations on the biology of *Parena lacticinta* Bates (Coleoptera : Carabidae), a predator of *Nephantis serinopa*. *Indian Coconut J.*, 9: 2–5.

Saha, S.K., Mukherjee, A.K. and Sengupta, T., 1992. Carabidae (Coleoptera : Insecta), of Calcutta. *Rec. Zool. Survey. India*, 1–63 pp.

Sathe, T.V. and Margaj, G.S., 2001. *Cotton Pests and Biocontrol Agents*. Daya Publishing House, pp. 1–161.

Sathe, T.V. and Bhosale, Y.A., 2001. *Insect Pest Predators*. Daya Publishing House, pp. 1–169.

Samal, P. and Mishra, B.C., 1978. *Casnoidea indica* (Thunb.) Carabid ground bettle predating on brown plant hopper, *Nilaparvata lugens*. (Stal.) of rice. *Curr. Sci.*, 47: 688–689.

Chapter 6

ROLE OF CHALCIDS IN CONTROL OF INSECT PESTS

T.V. Sathe[1] and A.M. Bhosale[2]

[1]*Department of Zoology,*
[2]*Department of Agrochemicals and Pest Management,*
Shivaji University, Kolhapur – 416 004, Maharashtra, India

Chalcid wasps (Figure 6.1) are good biocontrol agents of various insect pests. They belong to the insect order Hymenoptera, and contain 22,000 known species in the world. The estimated total diversity ranging from 60,000 to more than 5,00,000 species and many more have yet to be discovered and described.

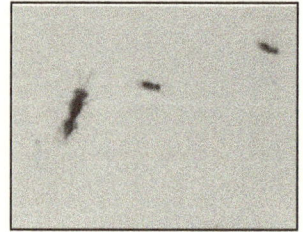

Figure 6.1: Chalcids

Most of the species are parasitoids of other insects, attacking the egg or larval stage of pest insects. The pests attacked by chalcids are found in at least 12 different insect orders including Lepidoptera (butterflies and moths), Diptera (true flies), Coleoptera (beetles), Hemiptera (true bugs), and other Hymenoptera, as well as two orders of Arachnida, and even one family of Nematoda.

Some species of chalcids are phytophagous and the larvae feed inside seeds, stems and galls. Generally beneficial to humans as a group, chalcids keep various crop pests under control, and many species have been imported to control insect pests.

Chalcids found on flowers, foliage and in leaf litter, but are very small sized, including the smallest of all known insects, *Dicopomorpha echmepterygis*.

They are tiny, dark-coloured wasps, often metallic blue or green with complex sculpturing on the body and reduced wing venation, They have a significant role to play in ecosystems as biocontrol agents of pest insects. Chalcids belong to the family – Chalcididae.

Chalcididae is a moderate-sized family within the Chalcidoidea, composed mostly of parasitoids and a few hyperparasitoids. The family is apparently polyphyletic and contain over 85 genera and over 1455 species worldwide. They are often black with yellow, red, or white markings, rarely brilliantly metallic, with a robust mesosoma and very strong sculpturing. The hind femora are often greatly enlarged, with a row of teeth or serrations along the lower margin.

Although chalcids are good biocontrol agents sometimes they acts as hyperparasites and becomes limiting factor for biocontrol agents.

Review of literature indicates that Indian Chalcids are rarely attempted except the work of Narendran and few others. Therefore, it is very interesting to know the role of chalcids in biological pest control as part of ecofriendly pest control.

According to Greathead, (1986) about 80 chalcid species are known to be pests of agriculture (mostly seed-feeders in the families Eurytomidae and Torymidae) and some chalcids are considered harmful because they are hyperparasitoids, but most are economically and environmentally beneficial. The large majority of chalcid species are primary parasitoids of other insects and arachnids. Over 800 chalcid species have been associated with targeted biological control programs. This represents about two-thirds of all biocontrol programs involving Hymenoptera, and about one-third of all biocontrol programs in which partial or complete economic control of an insect pest was achieved.

Biology and ecology of chalcids have been reviewed by Bendel-Janssen (1977). Grissell and Schauff (1997) states that the host range of chalcids is thought to exceed that of all other insect groups except for the order Diptera. Members are known to attack hosts in about 340 families of 15 insect orders (Blattaria, Coleoptera, Diptera, Hemiptera, Homoptera, Hymenoptera, Lepidoptera, Mantodea, Neuroptera, Odonata, Orthoptera, Psocoptera, Siphonaptera, Strepsiptera and Thysanoptera), as well as egg sacs of spiders (Araneae), ticks and gall-forming mites (Acarina), cocoons of pseudoscorpions (Pseudoscorpiones), and gall-forming Anguinidae (Nematoda). Chalcid parasitism occurred on all host life stages from egg to adult, as internal or external parasitoids, as primary or hyperparasitoids, and with their eggs laid in, on or away from the host (Grissell and Schauff, 1997). Some chalcids of the family Aphelinidae even parasitize the opposite sex of their own species (Woolley, 1997).

Like other Hymenoptera, most chalcids have a haploid-diploid mechanism of sex determination in which fertilized (diploid) eggs normally develop into females and unfertilized (haploid) eggs normally develop into males (arrhentokous development). However, males are unknown or are very rare for some species and in these species females produce females asexually (thelytokous development).

General Biology

Chalcidoids probably have a greater range of biological diversity than species of any other parasitic families. Parasitoid biology reaches its most elaborate development in the Chalcidoidea. There are solitary and gregarious species; ectoparasitoids and endoparasitoids; primary, secondary and tertiary parasitoids; polyembryonic species; and species with planidial larvae. Some species are extremely polyphagous and others may be very pest specific. All stages of pests are attacked, from the egg Mymaridae, Trichogrammatidae, Eulophidae, Encyrtidae, Aphelinidae, etc.) and to the pupa (Pteromalidae). Homoptera (especially Coccoidea and Aphidoidea) are attacked by Encyrtidae and Aphelinidae as egg, nymph or adult parasitoids. The eggs of Psocoptera are parasitized by mymarids and the eggs of Thysanoptera by Trichogrammatids. Nymphal Thysanoptera are parasitized by some Eulophidae, and immature Acari are attacked by encyrtids. While, spider egg sacs attacked by few species of Eulophidae, Encyrtidae and Pteromalidae.

Trichogrammatids (Figure 6.2) are also widely used in biological control. The chalcids have four different stages of life cycle *viz.* Egg, Larva, Pupa and Adult. The life cycle of *Brachymeria intermedia* an internal parasitoid of forest pest insect *Lymantria dispar* also contain above four stages.

Brachymeria bengalensis (Cameron) and *B.lasus* (Walk.) are pupal parasitods of *Pieris brassicae*. The female parasitoid preferred only 2 to 3 days old pupa for parasitization. They usually deposited one egg in a host pupa. On an average a female laid 19-21 eggs during its life time. The females lived longer than males. Sex ratio was highly female biased (Sudersana Devi and Singh, 2006).

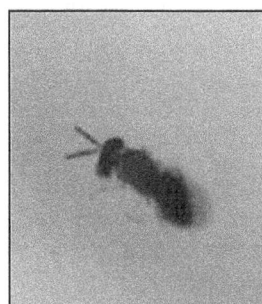

Figure 6.2:
Trichogramma

The biology of *Brachymeria bengalensis* (Cameron) and *B.lasus* (Walk.) was conducted under laboratory conditions (25 ± 1°C, 75 per cent RH, 12hr photoperiod). For initial culture, parasitized pupae of *Pieris brassicae* were collected from the field and kept under observations till the emergence of adult parasitoids. A pair of parasitoids was kept in the plastic container size 1 litre and was provided with 50 per cent honey solution soaked in cotton swab. The parasitoids were kept together for about 3-4 days before providing the pupae for oviposition. The newly formed pupae were inserted in to plastic container containing mated female parasitoid and parasitism was recorded. The parasitized pupae were separated and kept for further development of the parasitoid. The total duration of egg, larva, pre-pupa and pupa (endoparasite) of the parasitoid was noted. The adult longevity as well as sex ratio was also recorded. Number of parasitoids emerged from the pest insect was counted for considering the fecundity of single female parasitoid.

The pupal parasitoids *Brachymeria bengalensis* and *B.lasus* have showed four distinct life stages *i.e.* egg, larva, pupa and adult. From the egg to the pupal stage, they developed within the pest body and adult as the free living stage. These parasitoids are solitary in habit and occurred as a primary parasitoid. They are found to be specific in their preference for a particular stage of development of the pest species. The female parasitoid preferred only 2 to 3 days old pupae for oviposition.

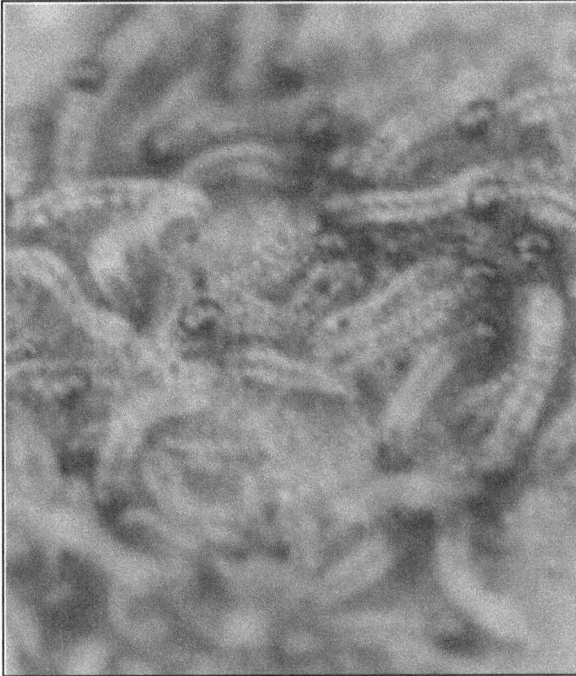

Figure 6.3: *Spodoptera* **Larvae.**

Fecundity

A female parasitoid usually deposits single egg in a host pupa. The adult *Brachymeria bengalensis* and *B. lasus* lays 18-21 eggs with a average of 19.5 eggs during its life span. According to Cherian and Basheer (1938) and Kamal (1938) *B. excarinata* (Pupal parasite of the leaf roller) and *B. femorata* (Pupal parasite of Cabbage white) laid a maximum of 230 and 180 eggs respectively. Whereas, Satpathy and Rao (1972) reported that in case of *B. nephantidis*, it laid on an average of 15-23 eggs. While Kapadia (1988) observed that fecundity of the female *B.excarinata* was on an average of 17-50 eggs in 30 days.

Development of Endoparasitic Stages

The duration of egg to adult emergence of the parasitoid *Brachymeria* sp. varied from species to species. The immature stage of *B. bengalensis* and *B.lasus* took about 28 days and 20 days, respectively. According to Satpathy and Rao (1972) the development period of immature stages of *B. nephantidis* was 16-18 days. While in *B.excarinata* the duration of egg to adult emerge varied from 17-18 days with an average of 13.53 days (Kapadia, 1988).

Adult Longevity

The adults of *B. bengalensis* survived for 14 to 25 days with an average of 17 days and 11 to 15 days with an average 13 days in *B.lasus* with 20 per cent sugar solution.

In case of *B.nephantidis*, the adult wasps lived for the longest period when fed with honey solution. But with other foods such as sugar and glucose solution the adult longevity were found to be reduced considerably (Satpathy and Rao, 1972). According to Joy and Joseph (1973) males and females of *B.nephantidis* lived for 24-60 days and 40-80 days respectively. Satpathy and Rao (1972), further reported the longevity of males and females to be about 7-8 days and 12-14 days, respectively. The total life cycle duration of *B.bengalensis* and *B.lasus* observed to be 40.5 and 33.5 days, respectively.

Sex Ratio

The sex ratio (male : female) was 1 : 2.05 in *B.bengalensis* and 1 : 2.17 in *B. lasus* favoring the females. Joy and Joseph (1973) also made similar observations in case of *B. nosatoi* whereas Kapadia (1988), reported higher male density of *B.excarinata* in the field condition. In all above cases the female chalcids never laid more than one egg in a single host pupa. From this, *Brachymeria* sp. can hardly be considered as prolific breeder in comparison to other related species. The parasitism of this pupal parasitoid in the field is dependent on the increase in the female population of the parasitoid.

Family : Chalcididae

Most chalcids (Figure 6.1) are solitary, primary endoparasitoids of pupae or mature larvae of Lepidoptera. There are also several that attack Diptera and Coleoptera and some that attack Hymenoptera and Neuroptera. Some species are facultatively hyperparasitic; and at least one genus is exclusively hyperparasitic. No phytophagous species known. More than 90 genera and 1500 species have been reported from the world.

Chalcidids are thus, predominantly solitary, primary endoparasitoids of Lepidoptera and Diptera, though a few species attack Hymenoptera, Coleoptera or Neuroptera; some tropical species are ectoparasitoids, and a few may be gregarious. Some species of chalcids are hyperparasitic. Some British species are endoparasitoids of Diptera, Coleoptera and Symphyta. Most are idiobionts, ovipositing into more or less fully grown hosts, such as mature larvae (in the case of parasitoids of Diptera), or young pupae (parasitoids of Lepidoptera). However, species of *Chalcis* are koinobiont parasitoids of *Stratiomys* (Diptera: Stratiomyidae). Some species oviposit into eggs of *Stratiomys*, which are laid in clusters on water-side vegetation (Cowan, 1979).

Female chalcids may lay about 200 eggs which are elongately-oval and may sometimes have a very short petiole. The first instar larva may be caudate or hymenopteriform, with or without spiracles, but with well-developed cuticular spines. Subsequent instars are mostly hymenopteriform (Dowden, 1935; Arthur, 1958). Pupation takes place in the host pupa. Most chalcidids overwinter as adult females, or as mature larvae in the hosts. *Brachymeria intermedia* is a parasitoid of *Lymantria dispar*, an introduced lepidopterous pest of a variety of trees in North America.

Rearing of Chalcids

Most immature and mature stages of the higher orders of insects (eg. Hemiptera, Coleoptera, Diptera, Lepidoptera) are attacked by chalcids in the field. Hence, stages of pests be collected from the field and reared in the laboratory for Chalcid emergence. The adult chalcids (Male and Female) are separated at emergence and later, used for mating purpose. After mating female chalcids are allowed to lay their eggs on immature/mature stages of pest insects for further development for mass production. Host culture be initially started in the laboratory. Chalcids are reared on respective stages of pest insects. The chalcid fauna can be reared easily (Boucek, *et al.*, 1981). Whatever hosts are collected it is best to put them in a suitable receptacle to await the emergence of the parasitoid(s), *e.g.*, glass tube with a cotton wool plug, brown paper bag, gelatin capsule, polythene bag, etc. Parasitized hosts can be distinguished from healthy hosts by their slightly different colouration or behaviour. Parasitized hosts may be darker than healthy ones or may move at a different pace or in a different way from healthy ones. The emerging insects are attracted to the light and collected into a glass tube.

Disadvantages of the Emergence Box

1. A fairly large proportion of the parasitoids may not find their way into the collecting tubes.
2. There may be several possible hosts in a sample from which the parasitoids are reared. For handling parasitoids, test tubes or specimen tubes are used. The parasitoids may climb on invertedly placed test tube and parasitoid may be separated.

Chalcidoidea as Biological Control Agents

The Chalcidoidea is the most important successful group used in applied biological control. Over 800 different species are associated with biocontrol programmes in one way or another. The families Aphelinidae and Encyrtidae, have proven extremely successful in the biological control of insects pests, although species of most other chalcidoid families have also been successfully utilized. The family Chalcidae is very prominent in biocontrol agents of insect pests.

The genus *Encarsia* from Aphelinidae, is one of the most important parasitoid groups exploited for biological control. Many species of this genus are used in the biological control of whiteflies (Aleyrodidae) and armoured scale insects (Diaspididae). Examples include *E. perplexa* Huang and Polaszek against citrus blackfly (*Aleurocanthus woglumi*) in the southern US and Caribbean (Clausen 1978), *E. smithi* (Silvestri) against spiny blackfly *Aleurocanthus spiniferus* (Kuwana 1934) and *E. inaron* (Walker) against ash whitefly in California. The best-known example of an *Encarsia* species, and probably the most well-known parasitoid used for biocontrol world-wide, is *E. formosa*. It is in use for 80 years against the greenhouse whitefly *Trialeurodes vaporariorum* and widely available commercially. *Encarsia* species are successfully used against scale insects, *E. berlesei* for white peach scale *Pseudaulacaspis pentagona*, and *E. perniciosi* for San Jose scale, *Quadraspidoitus perniciosus*. Aphelinid genera that include a number of species also have been used for classical biocontrol purposes, are *Eretmoceras* for control of whiteflies and *Aphytis* for the control of diaspidids.

Encyrtids (Encyrtidae) are also widely used in classical biological control programmes throughout the world, but most successfully in warmer climates. Mealybugs are controlled by using encyrtids as natural enemies. Perhaps the best known of these is the recently extremely successfully used *Anagyrus lopezi* in South America for the control of cassava mealybug (*Phenacoccus manihoti*). A species that has been used successfully for biocontrol purposes in Europe is *Psyllaephagus pilosus*. Since 1993, it has been released in Ireland, Wales, France and California (USA) for the control of eucalyptus psyllid. It successfully controlled the pest in these countries and has more recently been introduced into a South America, Peru and Argentina.

How to Collect Chalcids

For biological and taxonomical studies collection of chalcids is must. Following are the methods of chalcid collection.

1. Sweep Net and Sweeping

For the taxonomic studies sweeping is probably the most important way of collecting chalcids, since a relatively good diversity of species can be collected in short time. The material collected is temporarily stored into alcohol. The net contain aluminium rod and triangular ring like head which increases the surface area in contact with the ground at sweeping grassland or it allows easier penetration into dense vegetation.

The long handle allows the net to be used far away from the body, making sweeping underneath low, overhanging bushes easier, and also extends the area covered by each individual sweep. The netbag ring provided with durable material which allows the easy passage of air (*e.g.* silkscreen), but has a small enough mesh to prevent the escape of smaller specimens. The material for the netbag should be as translucent as possible. The netbag should not be deeper than arm's length (*i.e.* about 60cm) and should have a well-rounded bottom. A net back should not too long flap around during sweeping and damage the insects inside, whilst one that is too short cannot be folded over the frame to prevent escaping when the net is not in use. One that is too tapered will make removal of specimens more difficult.

The net can be adapted to allow a removable screen in order to screen out larger debris such as leaves. The size of the mesh can vary from about 4-10mm. The 0.4mm mesh will allow insects up to 7mm long to pass through.

Five minute sweeping bushes, trees or other vegetation is the best way for good catches. Sweeping for longer or too vigorously will result in damage to the insects in the net. Hence, the net should be emptied for several times.

2. Using a Light Box

A light box of size about 50 x 30 x 30 cm be constructed entirely of perspex or some other transparent material and used.

Use the box in a darkened area with strong light coming from one direction preferably with a pale background or that small insects can be seen easily. Emerging insects will be attracted towards the stronger light and can be collected with an aspirator as they walk up the walls of the box.

3. Using a Separation Bag

A separation bag be used as the same principle as a light box (Masner and Gibson, 1979). It consists of tripod holding a 30cm diameter circular frame that holds a strong canvas bag, the top of which is closed off by a circular perspex or glass lid. A hole about 12-15cm long is cut in the side of the bag about half way up. This is closed by a zip. The contents of the sweep net bag be emptied into the canvas bag and after a few minutes the emerging insects can be collected off the side of the bag.

4. Collection Direct into Alcohol

This is efficient method of collecting chalcidoids especially important for smaller specimens. Using this method the entire contents of the net are dumped into a polythene bag/plastic container/glass container/specimen tubes etc. containing 70 per cent alcohol. The escape of insects when transferring into a polythene bag, start with one that contains no alcohol. Dampen the sides of the polythene bag by spraying with a little alcohol then vacating the contents of net by everting the end of it into the polythene bag is good practice.

5. Beating

Holding a beating tray under a bush, branch or other suitable habitat, and then sharply hitting the branch with a heavy stick to dislodge insects is good method. Insects will fall directly into the tray and can be collected using an aspirator or fine paint brush dampened with alcohol.

Advantages

1. The catches are smaller,
2. many insects escape,
3. can only be used if the vegetation is reasonable high. However, beating is best done in cool weather or early or late in the day when insects are least active.

6. Pyrethrum Spraying

Insects are obtained from dead or rotten wood or from habitats unsuitable for sampling by other means. Pyrethrum spraying method is used to collect insects from small bushes or single branches. A polythene sheet is spread under a piece of vegetation or rotten wood which is then sprayed with a pyrethroid. Dying insects, as well as those falling out of crevices or holes in bark, etc. will fall into the sheet and can be collected using an aspirator, fine brush or forceps.

Disadvantages

1. Smaller individuals may get trapped in minute droplets of spray on leaves, bark, etc.
2. Specimens may become stiff and difficult to relax, which hinders mounting.

7. Canopy Fogging

This aerosol spraying method is especially successful for sampling taller trees or the canopy of tropical forests. It is one of the most productive methods of collecting many groups of chalcidoids, *e.g.* Eupelmidae, Encyrtidae, Aphelinidae, Signiphoridae and Trichogrammatidae.

8. Malaise Trap

The use of Malaise traps for collecting chalcidoids has probably revolutionised the study of chalcidoids.

A Malaise trap is used to collect large number of flying and occasionally flightless insects. The mesh of the net must be fine enough to prevent smaller chalcidoids from passing through easily. A Malaise trap is giraffe shaped, correctly constructed and sited, will provide a representative sample of chalcids from expected area and will collect even the smallest insects, *e.g. Megaphragma* spp. The trap may be run dry with an insecticidal fumigant in the collecting pot. 70-80 per cent alcohol is used as killing agent, the size of the catch will be greatly increased due to attraction to the alcohol itself.

Advantages

1. It need only be visited once every week or two weeks for emptying.
2. It can be serviced by a non entomologist.

9. Flight Intercept Trap

The trap consists of a 1m high, 2-3m wide length of black terylene netting slung vertically between two trees or posts. It is protected from rain by a transparent polythene roof that extends about 0.5m either side of the vertical netting for at least its length. A number of pans are placed beneath the netting for the whole of its length. The pans contain water with a drop or two of detergent, a saturated salt solution, a 50/50 ethylene glycol/water mix or some other suitable collecting medium. Flying insects hit the vertical net and fall into the pans. They can then be collected at regular intervals using the same method described below for yellow pan traps. The catch can be increased by spraying the netting with a long-lived contact insecticide.

10. Yellow Pan or Moericke Trap

This is an excellent method of collecting chalcidoids, notably mymarids and encyrtids, as well as other groups of insects. Species that are rarely swept or collected in Malaise traps can often be collected using this technique. Many insects are attracted to yellow colour. Hence, the trap consists of a shallow tray, about 60-75 mm deep and with an area of about 300-400 sq cm. This is painted bright yellow on the inside. The tray is placed on the ground in a suitable habitat such as grassland, a forest trail or clearing, etc. It is filled with water, saturated salt solution, or a 50/50 mix of ethylene glycol and water. If water only is used then the pan must be emptied at least once a day, ethylene glycol/water mix can run for about a week without being emptied.

11. Suction of Vacuum Sampler

A suction sampler is normally powered by a two-stroke motor that draws air through a length of tube about 5-30 cm in diameter. The open end of the tube is pushed into vegetation. Insects sucked in are trapped by a strong, fine net fitted across the inside of the tube within a few inches of the open end.

Disadvantages

1. Most of the larger species and many of the smaller ones tend to escape.
2. Apparatus is usually heavy and cumbersome to use.

12. Suction Trap

This is a reasonable method of collecting chalcids. The suction trap consists of a cylinder of 60cm diameter or so with a fine mesh gauze fitted inside. This leads into a collecting tube or chamber at the bottom containing saturated picric acid, water or alcohol. Air is drawn through the funnel by a fan. Passing insects are sucked down the gauze funnel and eventually fall into the collecting tube which must be emptied regularly.

Disadvantages

1. Needs an electric power source.

13. Pitfall Trap

The pitfall trap consists of a jar or another suitable receptacle sunk in the ground and partly filled with saturated picric acid solution. This can be left for a week without servicing. The specimens collected in the pitfall trap must be washed thoroughly in clean water before transferring to 70 per cent alcohol.

14. Light Trapping

This is perhaps one of the most overlooked methods for collecting chalcidoids. It is particularly efficient for collecting fig wasps in or near tropical forest habitats and also can collect species that are not known to fly at night including chalcid parasitoids.

References

Barnard, P.C. (Ed.), 1999. *Identifying British Insects and Arachnids: An Annotated Bibliography of Keyworks*. Cambridge University Press, pp. xii+353.

Boucek, Z., 1988. *Australasian Chalcidoidea (Hymenoptera): A Biosystematic Revision of Genera of Fourteen Families, with a Reclassification of Species*. CAB International, Wallingford, Oxon, U.K., Cambrian News Ltd; Aberystwyth, Wales, pp. 832.

Gauld, I.D. and Bolton, B. (Eds.), 1988. *The Hymenoptera*. Oxford University Press, Oxford, UK (Reprinted and revised, 1996), pp. xi+32.

Gibson, G.A.P., Huber, J.T. and Woolley, J.B. (Eds.), 1993. *Annotated Keys to the Genera of Nearctic Chalcidoidea (Hymenoptera)*. National Research Council of Canada, Ottawa, Canada, pp. xi+794.

Goulet, H. and Huber, J.T. (Eds.), 1993. *Hymenoptera of the World: An Identification Guide to Families.* Research Branch, Agriculture Canada, pp. vii+668.

Hanson, P. and Gauld, I.D., 1995. *The Hymenoptera of Costa Rica.* Oxford University Press, Oxford, UK, pp. xx+893.

Noyes, J.S., 1998. *Catalogue of Chalcidoidea of the World.* CD–ROM Series, ETI, Amsterdam, Netherlands.

Noyes, J.S. and Valentine, E.W., 1989. Chalcidoidea (Insecta: Hymenoptera) – Introduction, and review of genera in smaller families. *Fauna of New Zealand*, 18: 1–91.

Peck, O., Boucek, Z. and Hoffer, A., 1964. Keys to the Chalcidoidea of Czechoslovakia (Insecta: Hymenoptera). *Memoirs of the Entomological Society of Canada*, 34: 170, 289 figs.

Sharma, B.R., 1988. *Keys to the Insects of the European Part of the USSR, Volume III: Hymenoptera Part II.* Oxonian Press Pvt. Ltd., New Delhi, India.

Trjapitzin, V.A., 1978. *Oprediteli Nasekomikh Evreopeyskoy Chasti SSR. Tom III. Pereponchatokriliye Vtoraya Chasti.* Nauka, Leningrad, USSR (English Translation), pp. 759.

Chapter 7

ROLE OF HEMIPTEROUS PREDATORS IN INSECT PEST MANAGEMENT

T.V. Sathe*

*Department of Zoology, Shivaji University,
Kolhapur – 416 004, Maharashtra, India*

Introduction

The order Hemiptera contain more than 65,000 species in the word. By taking into the account of texture of wings the order is further subdivided into Heteroptera and Homoptera. Out of which the suborder Heteroptera plays important role by producing predatory species which has great importance in biological pest control programmes. Heteropteran bugs are minute to large, terrestrial or secondarily aquatic and are predatory in nature. Important predaceous families of the sub order Heteroptera refers to Pentatomidae, Coreidae, Lygaeidae, Pyrrhocoridae, Reduviidae, Ploiariidae, Notonectidae, Nabidae, Anthocoridae, Gerridae, Nepidae, Belostomatidae, Corixidae etc. Entomophagus predators are following types:

1. Predators which use some resin coated tibiae as traps.
2. Wait and grab predators
3. Pin and stalk predators
4. Run and pounce predators

* E-mail: profdrtvsathe@rediffmail.com

The review of literature indicates that Sahayaraj (2004) worked on biocontrol potential of Reduviids. Ambrose (2001) studied the role of Assissin bugs in Integrated pest management. Capriles (1990) provided systematic catalogue of the Reduviidae of the world. Cherian and Brahmachari (1941) reported three hemipteran bugs from South India for insect pest control. Lakkundi and Parshad (1987) developed mass multiplication technique for reduviids. Miles (1972) studied saliva of Hemiptera. Sahayaraj (1994) illustrated biocontrol potential of a reduviid *Rhinocoris marginatus* (Fab.) against a polyphagus pest *Spodoptera litura* (Fab.). Sahayaraj (1997) also developed laboratory rearing techniques for some reduviids. Sahayaraj and Abrose (1997) studied biocontrol potential of *Acanthaspis pedestris* Stal against gram pod borer, *Helicoverpa armigera* (Hubn.). Singh (1985) very first reported *Rhinocoris fuscipes* Fab. as predator for *Dicladispa armigera* (Oliver) while, Vanderplank (1958) reported a assassin bug *Platymeris rhadamanthus* Gerst as useful predator for Rhinocerous beetles *Oryctes boas* Fab. and *Oryctes moneros* (Oliv.). Cohen (1985) investigated simple method for rearing of a predatory bug *Geocoris ochropterus* by providing pupae of ant *Oecophylla smaragdina*. Mukhopadhyay (1997) reported crop association of a Geocorine predator in India and its biocontrol potential. Waddill and Shepard (1974) tested biocontrol potential of *Geocoris punctipes* and *Nabis* spp. against Epilachna beetle *Epilachna varivestis*.

Anthocorid predators and their biocontrol potential have been investigated by several workers. Noteworthy amongst them refer to Russel (1970), Parajulee and Phillips (1992), Ananthakrishnan and Sureshkumar (1985) etc.

Role of Hemipterans in Pest Control

Anthocorids and Pest Management

Anthocorid bugs belongs to the family Anthocoridae of order Hemiptera which contains about 650 species. The anthocorid bugs are also called as flower bugs or pirate bugs. Anthocorids feed on many pest insects and mite species.

Voracity of predators, rate of reproduction and synchronization of predator to preys are important factors which counts the efficacy of predator potential. Anthocorids which counts pest populations are listed in Table 7.1.

In India, *Orius maxidentex, O. tantilus, O. indicus* and *Carayonocoris indicus* are successfully utilized for control of thrips (Anthakrishanan and Sureshkumar, 1985). *Leptothrips karnyi* has been found highly predated by *Montandoniola moraguesi*. Eggs of thrips were highly predated by first and second instars of *M. moraguesi*. According to Nassar and Abdurahiman (1998), *Cardiastethus oxiguss* is very effective biocontrol agent of a coconut pest black headed caterpillar *Opisina arenosilla*. The predator attacks pupal and early stages of the pest insect. According to Jacobson (1991) *Anthocoris nemorum* successfully suppressed the population of *Frankliniella occidentalis* on cucumber crop in green house. Nagi (1991) assessed predatory potential of *Orius* species against 3 insect pests. He found order of feeding preference, *Thrips palmi* > *Thrips kanzawai* > *Aphis gossypii*. Finally, he concluded that *Orius* spp. has good bicontrol potential against above said pests. *Megalourothrips nigricornis* is pest of Red gram *Cajanus cajan* which infest flowers. According to Rajasekhara and Chatterji

Table 7.1: Hemipteran Pest Predators.

Sl.No.	Predator	Family	Insect Pest	Stage of Attack
1.	Lyctocoris beneficus (Hinra)	Anthocoridae	Chilo suppressalis (Walker)	Larva
2.	Orius tentius (Motsch)	Anthocoridae	Pectinophora gossypiella (Saunders)	Larva
3.	Andrallus spinidens Fab.	Pentatomidae	Helicoverpa armigera (Hubr.) Scirpophaga incertulus (Walk) Mythimna spp., Parnara mathias (Fab.), P. naso (Fab.) Sesamia inferens (Walk)	Larva
4.	Canthecona furcellata (Wolff.)	Pentatomidae	Earias spp., Parnara mathias (Fab.), Sesamia inferens (Walk.), Many caterpillars	Larva Larva
	Amyotea malbarica (Fab.)	Pentatomidae	Parnara mathias, (Fab.), P. naso (Fab.), Euproctis xanthorrhoea (Kollar) Sesamia inferens (Walk.) Mythimna spp. Malantis ismene (Crarmb.) Chilo sp., Spodoptera spp., S. litura (Fab.) S. mauritia (Bois) Scirpophaga incertulus (Walk.)	Egg + larva Egg + larva Egg + larva Egg + larva Egg + larva Egg + larva Egg + larva Egg + larva Egg + larva Egg + larva
	Pygomenida bangalensis (West wood)	Pentatomidae	Dicladispa armigera (Oliv.)	Nymph
	Pentatomid	Pentatomidae	H. armigera	Egg
5.	Geocoris tricolor (Fab.)	Lygaeidae	Trialeurodes ricini (Misra), Coccids, weevils, aphids, Pseudococcus nipae	Nymph Nymph Nymph
6.	Antilochus coqueberti Fab.	Pyrrhocoridae	Dysdercus koenigii (Fab.)	Nymph
7.	Anthocoris nemaralis	Anthocoridae	Psylla pyricola	Nymph

Contd...

Table 7.1–*Contd...*

Sl.No.	Predator	Family	Insect Pest	Stage of Attack
8.	*Orius niger* (Wolf)	Anthocoridae	*Frankliniella occidentalis*	Nymph
			Thrips tabaci	Nymph, Adult
9.	*Orius maxidentex* Ghauri	Anthocoridae	Aphids, thrips	Nymph
10.	*Orius albidipennis*	Anthocoridae	Aphids, Thrips	Nymph, Adult
			P. gossypiella	Larva
			Earias insulana	Larva
			Helicoverpa armigera	Larva
11.	*Orius indicus*	Anthocoridae	*Megalourothrips nigricornis*	Nymph, Adult
12.	*Orius* sp.	Anthocoridae	*Bemisia tabaci*	Egg, Nymph
13.	*Anthocoris minki*	Anthocoridae	*Taeniothrips rhopalantennalis*	Nymph
			Aphis pomi	N + A
14.	*Orius insidiosus*	Anthocoridae	*Pectinophora*	Larva
			Earias	Larva
			H. armigera	Larva
15.	*Montandoniola moruguesi* (Puton)	Anthocoridae	*Gynaikothrips ficorum*	Nymph
			Androthrips flavipes	Nymph
			Liothrips karnyi	Nymph
	Xylocoris clarus	Anthocoridae	Thrips	Nymph
16.	*Harpactor costalis* Rev.	Reduviidae	*Dysdercus koenigii*	Nymph, Adult
17.	*Coranus apinicutis* Reuter	Reduviidae	*Helicoverpa armigera*	Larva
18.	*Oncocephalus annulipes* Stall.	Reduviidae	*H. armigera*	Larva
19.	*Rhinocorus kumarii*	Reduviidae	Red cotton bug	N + A
20.	*Rhinocorus fuscipes* Fab.	Reduviidae	*H. armigera*	Larva
21.	*R. marginatus* Fab.	Reduviidae	*Spodoptera litura*	Larva
			Red cotton bug *D. koenigii*	Nymph

Contd...

Table 7.1–Contd...

Sl.No.	Predator	Family	Insect Pest	Stage of Attack
22.	Sycannus indagator Stal.	Reduviidae	H. armigera	Larva
23.	Sphedanolestes aurescens Distant	Reduviidae	Nephantis serinopa	Larva
24.	Ectomoloris tibialis Distant	Reduviidae	Red cotton bug, D. koenigii	Nymph and Adult
25.	Catamiarus brevipennis	Reduviidae	Red cotton bug, D. koenigii	Nymph and Adult
26.	R. kumarii	Reduviidae	H. armigera	Larva
			S. litura	Larva
			Euproctis mollifera	Larva
27.	R. marginatus	Reduviidae	S. litura	Larva
			D. koenigii	Nymph
			M. pustulata	
28.	Platymeris laevicollis Distant	Reduviidae	Oryctes spp.	Larva
29.	Anisups bouvieri	Notonectidae	Mosquitoes	Larva
	Buevoa sp.			
30.	Ranatra sp.	Nepidae	Nepidae	Aquatic insects
	Nepa cinerea	Nepidae	Nepidae	Aquatic insects
	Laceotrephes	Nepidae	Nepidae	Aquatic insects
31.	Gerris tristan	Gerridae	Nilaparvata lugens	Nymph +Adult
	Limnogonus sp.	Gerridae	Jassids	
	L. nitidus (M.)	Gerridae	Jassids	
32.	Psallus sp.	Miridae	Thrips	Nymph + Adult
33.	Tytthus parviceps (Reut.)	Miridae	Brown plant hopper	Nymph
	Cyrtorhynus lividipennis (Rout.)	Miridae	Brown plant hopper	Nymph
			Leaf hopper	Nymph
			White backed plant hopper	Nymph

Contd...

Table 7.1–*Contd...*

Sl.No.	Predator	Family	Insect Pest	Stage of Attack
34.	Geocoris ochropterus (Fieber)	Lygaeidae	**Aphids:** *Aphis gossypii* G.	Nymph + Adult
			A. craccivora (Koch.)	Nymph + Adult
			Rhopalosiphum maidis (Fitch)	Nymph + Adult
			Myzus persicae Sulzer	Nymph + Adult
			Aphis traveresi (Del G.)	Nymph + Adult
			Toxoptera sp.	Nymph + Adult
			Jassids: *Amrasca binotata* Druchi	
			A. devastans Dist.	
			A. kerri Pruthi	
			Thrips: *Scirtothrips dorsalis* Hood	Nymph
			Taeniothrips distalis	Nymph
			Caterpillars: *Euproctis latisfascia* Walk.	Larva
			Sylepta derogata Fab.	Larva
			Helicoverpa armigera (Hubn.)	Larva
			Weevil: *Myllocerus blandus* Fst.	Eggs, Larva
			Coccid: *Centrococcus insolitus* Gr.	Nymph+A
35.	*G. aligerhansis*	Lygaeidae	Aphids, jassids, coccids	Nymph + Adult
	G. jucundus		Aphids, jassids	Nymph + Adult
36.	*Stenonabis tagalicus* (Stal.)	Nabidae	White backed plant hopper	Nymph + Adult
	Tropiconabis capsiformis	Nabidae	Green leaf hopper	Nymph + Adult
	Nabis sp.	Nabidae	**Jassids:** Plant bugs, *Heliothis* spp.	Nymph + Larva
37.	*Microvellia* sp.	Vellidae	Brown plant hopper	Egg, Nymph
	M. sytolinrsys (Berg)		Jassids	
38.	*Hydrometra* sp.	Hydrometridae	**Jassids:** Brown plant hopper,	Nymph
			White back plant hopper,	Nymph
			Green leaf hopper	
39.	*Belostoma indica*	Belostomatidae	Aquatic pests	Nymph/
	Lithocerus indicus	Belostomatidae	Aquatic pests	Larva
40.	*Corixa hieroglyphica*	Coreidae	Mosquiotes, Gnates	Larva
	Micronecta striata	Coreidae	Gnates	Larva

(1970) *Orius indicus* is very good biocontrol agent of *M. nigricornis*, both nymphs and adults of the predator feed on above pests on red gram crop.

Mass Rearing Technique for Anthocorid Bugs

Mass rearing techniques for some Anthocorid bugs have been developed by some workers (Van Lentern *et al.*, 1977; Schmidt *et al.*, 1995; Ballal *et al.*, 2003 etc.) According to Schmidt *et al.* (1995) mass rearing of *O. insidiosus* is possible in "zip-lock" plastic bugs. For providing water to predatory bug, 2-3 cm sections of snap pean pods are placed in each bag. Side walls of the bags are kept separate by using V shaped index card at bottom of the bag. After mating bug allowed to lay eggs on card sheet and then the card sheet is placed in the bag for hatching eggs. Egg card and plant material do not deteriorate for longer period in plastic bags. Thus, after hatching eggs large number of nymphs can be collected rearing in the plastic bags for adult emergence. Oviposition occurred in the same container. Each bag facilitates for 200-600 individual predators. For maintaining 10,000 including 5000 adults 4-5 h are needed per week.

Mass Rearing of *C. exignus*

As food for predator, eggs of *Corcyra cephalonica* are used. *C. cephalonica* can be reared by the method proposed by Sathe (2004). About 40 and 300 eggs are required for predator feeding purpose for nymphal and adult stage respectively. Per day consumption rate was 3-4 and 10-11 individuals for nymphal and adult stage respectively. Incubation period and nymphal periods were 3.50 days and 17 days respectively. Adult longevity of males was 35.75 days and of female 66.00 days. Predator can produce its progeny with high rate upto 25 days.

Geocorid Bugs in Pest Control

Geocorid bugs belongs to the family Geocoridae of order Hemiptera. Good number of genera are known to the science but the genus *Geocoris* constitutes largest number of the family. The genus *Geocoris* contain about 125 species. About 25 species of Geocorids have been reported from diverse agroecosystems of Indian subcontinent. Important species of Geocorids which act as biocontrol agent are listed in Table 7.1 with scientific name, preys and stage of pest they attack. Geocorid bugs feed on small insect pest species found on various crops such as cotton, lady's fingure, maize, jowar, groundnut, sunflower, sesame, citrus, guava, potato, brinjal, soyabean, etc. According to Yadav (2001), *G. ochropterus, G. jucundus* and *G. aligerhansis* are found to be most abundant and effective predators for controlling pests such as, Aphids, jassids, thrips, caterpillars, weevils etc.

Mass Rearing of *G. ochropterus*

Mass rearing of *G. ochropterus* has been developed by Mukhopadhyay and Sanniarahi (1993). The above bug is reared on eggs and larvae of *Corcyra cephalonica*. However, for increasing weight and fecundity of predator, pupae of an ant *Oecophylla smaragdina* F. have been provided in addition to protein, carbohydrates and lipids as basic nutrients. Mating occurs in insect cage (size 30 x 30 x 25 cm). Mated female start egg laying. However, a company of male to female, leads to lay more eggs. Continuous supply of fresh food avoids cannibalism in the predator species. Therefore, ant pupae are supplemented periodically as per the need of predator. Ant pupae can be stored in refrigerator at -10°C and RH 62 per cent. The above said cage is made up of glass doors and Top except the one side is provided with nylon/cotton mesh sleeve for handling insects. Under laboratory conditions (27±1°C, 65-70 per cent RH, 12 hr photoperiod) predator can multiply satisfactory. Geocorids become phytophagus in scarcity of preys and get survived in the field condition. Thus, their releases in small scale or large scale are helpful in pest control.

Utility

According to Mukhopadhya and Das (2004) *G. ochropterus* consume about 15 aphids per day in the field and predator can survive for more than 21 days. The above predator also predates upon Tea thrips and suppress population in field conditions. According to Lingren *et al.* (1968) when *G. punctipes* released at 6,30,000 per acre in cotton field egg and larval population of *Heliothis virescens* was found reduced upto 88 per cent. This bug is equally effective against Mexican beetle in soyabean fields. *G. bullatus* and *G. pallens* can suppress the populations of green peach aphid and *Myzus persicae*. Geocorids also control *Helicoverpa armigera* pest by feeding upon its eggs. Geocorids have good role in pest control, therefore, they should be exploited on large scale in pest control techniques. Secondly, they are supposed to be safer for pollinators since geocorids attack either sessile colonical forms or slow moving immature forms of insect pests.

Reduviids in Pest Control

Reduviids are grouped under the family Reduviidae of order Hemiptera. Out of 450 species reported from Reduviidae at least 115 species are supposed be very good biocontrol agents of insects pests from agricultural ecosystems. From Tamil Nadu about 120 species of Reduviids have been reported. Records of Reduviids from several states including Maharashtra are not known. The Reduviid bugs are commonly called as Assassin bugs, wheel bugs, Ambush bugs, Thread legged bugs, Asian ladybugs etc. In Africa they are called as Hunting bugs. Reduviids or Assassin bugs are found in tropical and temperate countries of the world. Their diversity is peaked in tropics. Reduviids are usually found in concealed habitats such as underneath the stones, dried barks and in crevices of all types in scrub jungles, semiarid zones and tropical rain forests adjoining agroecosystems.

Reduviids are Characterized by:

1. Presence of modified front legs for grasping and holding prey.
2. Presence of tibial pads used for holding prey.
3. They have migratory habit for searching preys.
4. They consume more preys than their need
5. They complete only one generation during a single year.
6. Prey selection capacity is very good.
7. They are either black and red or yellow and black coloured.

Mass Rearing of Reduviid Bugs

Reduviids like *Acanthaspis siva* and *Rhinocoris marginatus* have been reared on small scale. Rearing is influenced by female parental age, decamouflaging, rearing substrata, type of preys and food, temperature, artificial diet etc. Reduviids can be reared by providing larvae of *Corcyra cephalonica*. However, webbing provided by larvae of *C. cephalonica* protect larvae from predation by reduviids. Therefore, refrigerated larvae are exposed to Reduviids as food. The life cycle of Reduviids completed within 132 days with above said refrigerated *C. cephalonica*. The above two predators can also be reared on preys such as *S. litura, E. vittella* and *C. cephalonica*. Larval card stripes are prepared by fixing 50 larvae on card stripes in different rows. The larvae then offered to the predators as food. 50 larvae per container are kept for rearing purpose for predators. Frozen larvae are also given to reduviids as another concept of food for predators. In addition, artificial diet is also available for rearing of reduviids under laboratory conditions. With above methods reduviids are mass reared and released in the field.

Utility

Rhinocoris marginatus is effective predator of *Aphis craccivora*. In cotton ecosystem reduviids suppress the population of red cotton bug *Dysdercus koenigii*. It has been reported that *Ectomoloris tibialis, Catamiarus brevipennis, Rhinocoris kumarii* and *R. marginatus* have controlled 50 per cent population of *D. koenigii* on cotton. As much as 84 per cent predation was caused by *E. tibialis* to red cotton bug. Both, adults and nymphs were predated in above case. Similarly, *H. armigera* larval population was controlled upto 50 per cent by *R. kumarii*. Likely tobacco caterpillar *Spodoptera litura* population was also effectively suppressed by *R. kumarii*.

In groundnut ecosystem *R. kumarii* predated *Aphis craccivora* with a very high potential. On the crop Lady's finger *H. armigera* is predated upto 65 per cent in Maharashtra.

In Maharashtra pigeon pea *Cajanus cajan* (Linn.) is attacked by the insect pests like *H. armigera* and Tussock caterpillar *Euproctis subnotata*. Both above pests are suppressed by *R. fusicies* at considerable level (34 per cent). According to Goel (1978) *T. indica* is good biocontrol agent for many paddy pests. *Exelastis atomosa* (Walsingham) also been suppressed upto 10 per cent by *R. fusicies* on red gram. When 100 adults of *R. kumarii* released in cotton ecosystem *D. singulatus* population found decreased by 6.00 per cent. Sahayaraj (1999) released 5000 bugs of *R. marginatus* 1 ha at 15 days interval for 3 times for control of *S. litura* and *H. armigera*. Very encouraging results have been reported against above pests. According to Sahayaraj (2004) when *R. kumarii* is released with 5000/ha against *S. litura* it will give 25 to 50 per cent control while, *H. armigera* population suppressed from 40 to 56 per cent in the field condition, particularly on groundnut crop. The same predator is also effective against *A. craccivora*. The predator reduced 16.5 per cent population of *A. craccivora*. Thus, *R. kumarii* is very good biocontrol agent of *S. litura, H. armigera* and *A. craccivora* in India and plays very important role in IPM. On groundnut crop when, *R. marginatus* is used along with other control measures as IPM, it increases yield of the crop and cost benefit ratio favouring benefit.

Integrated effects of *R. marginatus* have been tested by Sahayaraj (2004). He used the above predator for pest management with following combinations - Groundnut + vitex + Reduviid, Groundnut - Pongamia + Reduviid, Ground + Neem + Reduviid and Groundnut + Calotropis + Reduviid. In all cases cost benefit ratio was very much favouring benefits. Sahayaraj (2004) studied cost benefit ratios in some of the intercropped models along with predators. He noted higher cost benefit ratio from maize intercropping fields as compared to caster intercropping. Predators are increasingly adopted in IPM strategies since they are specific and safer to non target species, beneficial insects, higher animals and human being. *A. modicella* and *S. litura* were potentially controlled in maize and caster intercropping (Sahayaraj, 2004) by above mentioned predator.

Reduviid Limiting Factor

In the natural habitat reduviids are parasitized or predated by other species. The eggs of *R. albopunotatus* are parasitized by *Hadronotus abntestiae* Dodd, *Hadronotus* spp. and *Telenomus* sp. Similarly, eggs of *R. marginatus* are parasitized by *Trissolcus* sp. During the rainy season Reduviids are attacked by Bacterial diseases. Likely they are attacked by fungi like *Aspergillus flavus*. During rearing and field condition care of above factors be taken. Reduviids have ability to migrate from one place to other attempts should be made to avoid migration by providing proper food, shelter and mate. Before release of reduviids in field for control of pests density dependant factor should be investigated for better and appropriate use of predators. Pest abundance and stages of pest attacked by predator be investigated for appropriate predator release.

References

Ambrose, D.D., 2001. Assassin bugs (Heteroptera : Reduviidae) in integrated pest management programme : Success and strategies. In: *Strategies in Integrated Pest Management*, (Eds.) S. Ignacimuthu and Alok Sen. Phoenix Publishing House Pvt. Ltd., New Delhi, pp. 73–85.

Ananthakrishnan, T.N. and Sureshkumar, N., 1995. Anthocorids (Anthocoridae : Heteroptera) as efficient biocontrol agents of thrips (Thysanoptera : Insecta). *Curr. Sci.*, 54(19): 987–990.

Capriles, J.M., 1990. *Systematic Catalogue of the Reduviidae of the World (Insecta : Heteroptera)*. University of Puerto Rico (Mayaguez), pp. 694.

Lakkundi, N.H. and Parshad, B., 1987. A technique for mass multiplication of predator with sucking type of mouth parts with special reference to Reduviids. *J. Soil. Biol. and Ecol.*, 7: 65–69.

Mukhopadhyay, A., 1997. Crop association of Geocorine predator (Insecta : Hemiptera) in India and its biocontrol potential. *Proc. Zool. Soc.*, Calcutta, 50(1): 12–18.

Mukhopadhyay, A. and Sannigrahi, S., 1993. Rearing success of a polyphagus predator *Geocoris ochropterous* (Hemiptera : Lygaeidae) on preserved ant pupae of *Decophylla smaragdina*. *Entomophaga*, 32(2): 214–219.

Parajulee, M.N. and Phillips, T.W., 1992. Laboratory rearing and field observations of *Lyctocoris campestris* (Heteroptera : Anthocoridae), a predator of stored product insects, *Ann. Ent. Soc. Am.*, 85(6): 736–743.

Russel R.J., 1970. The effectiveness of *Anthocoris nemorum* and *A. confusus* (Hemiptera : Anthocoridae) as predators of the sycamore aphid *Drepanosiphum platanoides* 1; the number of aphids consumed during development. *Ent. Exp.* and *Appl.*, 13: 194–207.

Sathe, T.V. and Bhosale, Y.A., 2001. *Insect Pest Predators*, DPH, Delhi, pp. 1–169.

Sahayaraj, K., 1994. Biocontrol potential evaluation of the reduviid predator *Rhinocoris marginatus* (Fab.) to the serious groundnut pest *Spodoptera litura* (Fab.) by functional response study. *Fresenius Envir. Bull.*, 3: 546–550.

Sahayaraj, K., 1997. Laboratory rearing of predaceous bugs with special reference reduviids (Insecta : Hemiptera : Reduviidae). *Zoo's Print*, 13(5): 17–18.

Sahayaraj, K. and Ambrose, D.P., 1997. Biocontrol potential of *Acanthaspis pedestris* Stal (Insecta : Heteroptera : Reduviidae) to *Helicoverpa armigera* Hubner of bhendi. *Madras Agri. J.*, 84(5): 294–295.

Schmidt, J.M., Richards, P.C., Nadel, H. and Ferguson, G., 1995. A rearing method for the production of large numbers of the insidious flower bug, *Orius insidiosus* (Say) (Hemiptera : Anthocoridae). *Can. Entomol.*, 127: 445–447.

Singh, O.P., 1985. New Records of *Rhinocoris fuscipes* Fab. as a predator of *Dicladispa armigera* (Olive). *Agri. Sci. Digest*, 5: 179–180.

Van Lenteren, J.C., Ruskam, M.M. and Timmer, R., 1997. Commercial mass production and pricing of organisms for biological control of pests in Europe. *Biological control*, 10: 143–149.

Vanderplank, F.L., 1958. The assassin bug *Platymeris rhadamanthus* Gerst. (Hemoptera : Reduviidae) a useful predator of the *Rhinoceros* beetles *Oryctes boas* F. and *Oryctes moncros* (Oliv). *R. Ent. Soc. S. Africa*, 21: 309–314.

Waddil, V.H. and Shepard, B.M., 1974. Potential of *Geocoris puntipes* and *Nabis* spp. as predators of *Epilachna varivestis*. *Entomophaga*, 19: 421–426.

Chapter 8

GRASSHOPPERS IN PEST MANAGEMENT: BIOCONTROL POTENTIAL OF GRASSHOPPER *HIEROGLYPUS BANIAN* FROM PADDY ECOSYSTEMS OF KOLHAPUR REGION

T.V. Sathe, A.R. Bhusnar and Nilam Shendage

*Department of Zoology, Shivaji University,
Kolhapur – 416 004, Maharashtra, India*

ABSTRACT

Hieroglypus banian (Orthoptera : Acrididae)is serious pest of paddy in Kolhapur region. Their control with pesticides lead several problems like resistance in pests, pest resurgence, secondary pest outbreak, pollutions, health hazards, killing of beneficial organism, etc. Therefore, hoping biological control of *H. banian* biocontrol potential has been assessed with the help of natural enemies. Fungi, bacteria, nematodes, protozons, parasitoids, insect-predators and vertebrate predators are found feeding on grasshoppers in paddy ecosystems of Kolhapur region. The potential of natural enemies of grasshopper has been discussed in the paper.

Keywords: *Hieroglypus banian, Paddy grasshopper, Biocontrol potential, Natural enemies, Kolhapur region.*

Introduction

Grasshoppers (Orthoptera:Acrididae) constitute a large group of insects as serious and polyphagous pests of several agricultural, horticultural, floricultural and forest crops. Their control with pesticides lead serious problems such as air and water pollution, health hazards, killing of beneficial insects, pest resistance, pest resurgence, secondary pest outbreck, interruption to ecocycles, etc. The above facts indicate that there is need to find out alternative for chemical control. In view of these problems, the possibilities of biological control have been explored in various countries. A broad range of insect enemies, both parasitoids and predators, as well as nematodes and microbes have been used in pest control. In past, attempts on biological control of grasshoppers have been made by Greathead (1963), Rao *et al.* (1971), Rees (1973), Clausen (1976), Soper and Ward (1981), Nickle (1981), Canning (1981), Sankaran (1991), Sathe and Bhoje (2000), etc. Hoping biological control of *Hieroglyphus banian* as ecofriendly control, present work was carried out.

Material and Methods

Various stages of grasshoppers were collected from the field and their rearing was made under laboratory conditions (25±1°C, 70-75 per cent R.H.12 hrs photoperiod) by providing their natural food material and were screened for their natural enemies such as fungi, bacteria, nematode and protozoan parasites and insect parasitoids. Insect, invertebrate and vertebrate predators were recorded in the paddy ecosystems in Kolhapur district by one man one hour search method. The predators were collected and tested under laboratory conditions and their potential has been finalized. The natural enemies have been identified by consulting appropriate literature.

Results

1. Egg Parasites

During study period a solitary endoparasite *Scelio pembertoni* Timb. (Scelionidae) have been reported. The parasite completed it life cycle within 4-8 weeks. The parasite attacked diapausing as well as non diapausing eggs of grasshoppers. Chalcid parasitoids recorded on eggs of grasshopper refer to *Leefmansia* sp., and *Eupelmus* sp. and *Stichotrema* sp. About 10-22 per cent parasitism was recorded by these parasitoids.

2. Egg Predators

Mostly Bombuyliids *Anastoechus* sp. and *Systoechus* sp. were found feeding on eggs of *H. banian;* From Callophorids–*Stomorphina lunata* (Fab), from Coleoptera,

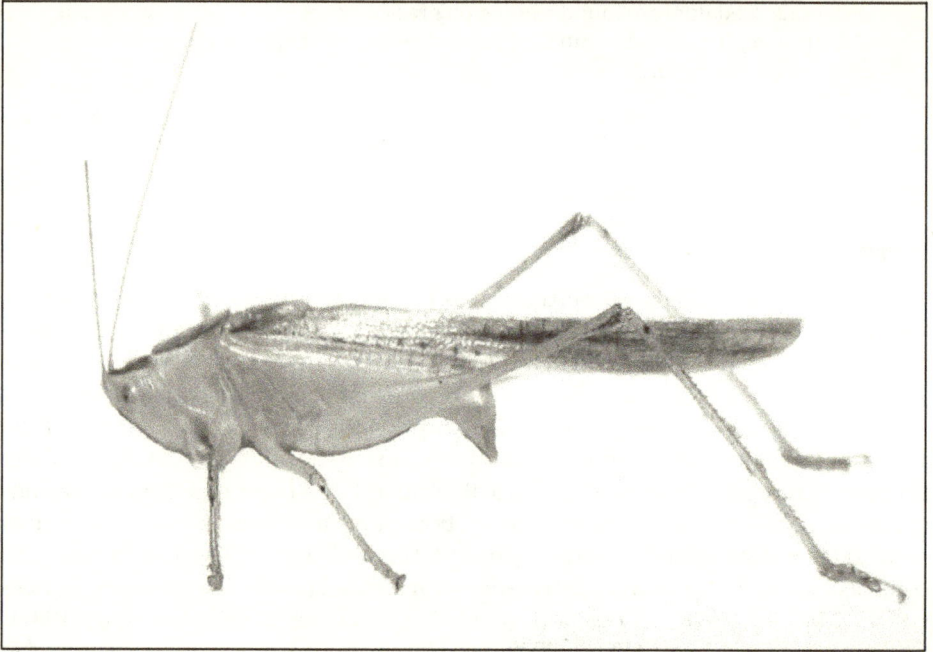

Figure 8.1: *Conocehalus insertus.*

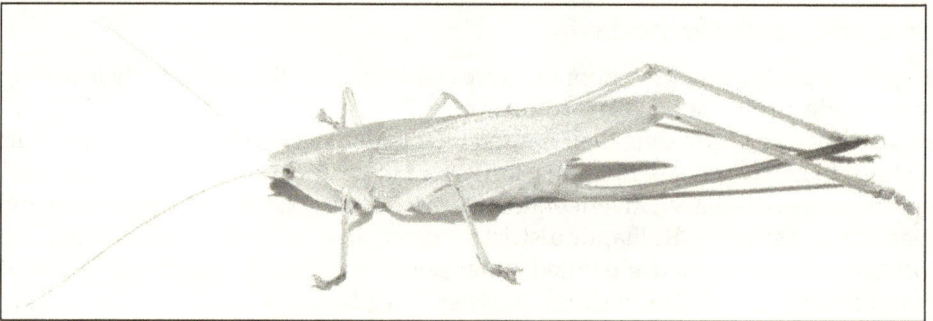

Figure 8.2: *Econocephalus incertus.*

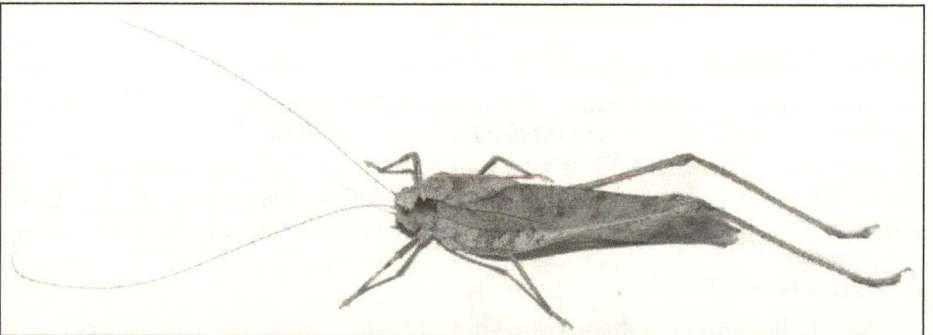

Figure 8.3: *Mecopoda elongata.*

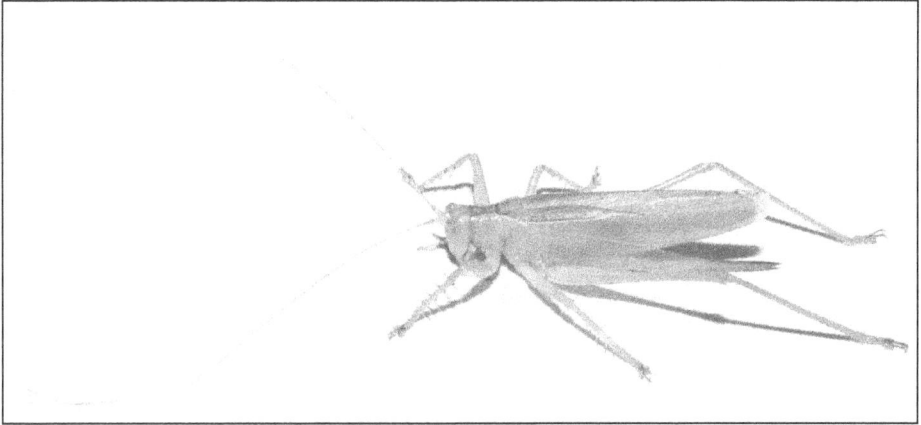

Figure 8.4: *Hexacentrus unicolar.*

Trogids, Clerids, Meloids were found feeding on eggs of grasshoppers. *Zonabia pustulata* attacked 30 per cent eggs of *H. banian*.

3. Nymph and Adult Parasitoids

Trichopsidea sp. and *Neorhynchocephala* sp. of the family Nemestridae of order Diptera have been parasitized grasshoppers. *Acridomyia* sp. (Muscoidae) was also found parasitizing eggs. From Nematoda *Mermis* sp. and *Agamermis* sp. attacked 15 per cent nymphs and adults.

4. Microbial Pathogens

Following pathogens attack 22 per cent population of *H. banian*.

Protozoa – Microsporidia –*Nosesma locustae*

Bacteria – *Coccobacillus* sp.

Fugi – *Entomopthora* sp.

Vertebrate Predators

About 15-35 per cent predation was noticed due to vertebrate predators.

1. Aves

Indian maynah – *Acridotheres tristis* (Linn.) - attacked 20 per cent

House sparrow – *Passer domesticus* (Linn.) - attacked 10-12 per cent

King crow – *Dicrurus adimillis* (Linn.) attacked 15-17 per cent

House crow – *Corvus splendens* Vie - attacked 5-10 per cent

Shikra – *Accipiter badius* - attacked 10-15 per cent

Predatory Grasshoppers

Grasshoppers also act as biocontrol agents. They feed on jassids, midges, small moths, mosquitoes and many small insects in paddy fields and at light source. Predatory grasshoppers are shown in Figure 8.1–8.4

Discussion

According to Sunkraran (1991) very few attempts have been made to use natural enemies to control grasshoppers an any large scale. Indian mynah bird *Acridotheres tristis* L. was introduced in Mauritius in 1762 for control of red locust *Nomadacris septemfasciata* (Serv.) for the first time in the world as first international movement of biological pest control. However, according to some workers reptiles, birds and other general predators are not introduced from one geographical area to another because of the possible adverse ecological impact on local fauna.

In Indonesia *Sixava nubile* (Stal) was controlled by using egg parasitoids *Leefmansia bicolar* Wtst. and *Doirania leefmansi* Wtst, above two parasitoids provided 85-95 per cent control of above coconut grasshopper.

In Hawaii *Oxya chinensis* was controlled by two species *Scelio*. In 1930-31. Similarlyin 1996 *Schistocerca vaga* Scudder was controlled by using a Sarcophagid *Blaesoxipha filipievi* from East Africa and *Pimelia inexpectata* (Sen.) from India.(Rao et al, 1971).

Moroccan locust *Dociostaurus maroccanus* Thnb. was controlled in Italy in 1946 by introducing a meloid beetle *Mylabris variabilis* Pall and two bombyliid flies *Cytherea obscura* Fab. and *Systoechus ctenopterus* Mik. In Pillippines in 1913 unsucessful attempt of control of locusts was made by introducing *Coccobacillus acridiorum*.

In the present study all natural enemies reported were more or less good biocontrol potential of *H. banian*. However, their mass multiplication and appropriate use need further trials. The basic studies on various aspects of natural enemies of grasshoppers will be helpful their mass rearing and exploitation as an alternative to chemical control and ecofriendly control in pest management stratergies.

Acknowledgement

Authors are thankful to Shivaji University, Kolhapur authority for providing facilities to authors.

References

Canning F.V., 1981. Insect control with protozoa. In: *Biological Control in Crop Production*, BARC Symposium, No. 5. Osmun Publishers.

Creathead, D.J., 1963. A review of insect natural enemies of Acridoidea (Orthoptera). *Trans. R. Ent. Soc., Lond.*, 114: 437–517.

Nickle, W.R., 1981. Nematodes with potential for biological control of insects and weeds. In: *Biological Control in Grasshopper*, BARC Symposium, 5(13): 181–199.

Rao V.P., Ghani, M.A., Sankaran T. and Mathur, K.C., 1971. A review of biological control of insects and other pests in south-east Asia and the pacific region. CIBC, *Tech. Comm.*, 6: 146.

Rees, N.E., 1973. Arthropod and nematode parasites, parasitoids and predators of Acrididae in America north of Mexico, USDA. *Tech. Bull.*, 14(60): 288.

Soper, R.S. and Ward, M.G., 1981. Production, formulation and application of fungi for insect control. BARC Symposium, 5(12): 161–180.

Sankaran, T., 1991. Biological control of grasshoppers: A status resume. *Hexapoda*, 3(1&2): 1–5.

Sathe, T.V. and Bhoje, P.M., 2000. *Biological Control of Insect Pests*.Daya Publishing House, pp. 1–121.

Chapter 9

LACEWINGS FOR BIOLOGICAL PEST CONTROL:

(A) Biodiversity of Lacewings (Order : Neuroptera) from Kolhapur Agroecosystems

T.V. Sathe and S.R. Sawant

Department of Zoology, Shivaji University,
Kolhapur – 416 004, Maharashtra, India

ABSTRACT

Lacewings (Order : Neuroptera) are exclusively predaceous on insect pests thus, act as biocontrol agents. Therefore, biodiversity of lacewings has been studied from agroecosystems of Kolhapur. In all 20 species of Chrysopids belonging to the genera *Chrysopa, Centameva, Chrysoperla, Glenochrysa, Mallada* and *Plesiochrysopa* have been recorded. The Chrysopids attacked hemipterous insects such as mealy bugs, aphids, jassids, scales, whiteflies, woolly aphids, caterpillars, thrips, etc. from various agroecosystems. Their occurrence has also been reported from Kolhapur region of Maharashtra.

Keywords: Lacewings, Neuroptera, Predators, Biocontrol agents, Diversity.

Introduction

Chrysopids belong to the Order Neuroptera. These are commonly known as green lacewings. The adults of many species feed on honey dew and pollens. The larvae are predaceous, feeding on eggs and neonate larvae of lepidopterans, nymphs and adults of whiteflies, aphids and other homopterous pest insects. They are also known as aphid lions and found mostly in agricultural and horticultural ecosystems including plantation crops (Jalali, 2003). From India about 70 species of Chrysopids have been recorded. Being predators, they are widely used in biological control programmes of insect pests. The review of literature indicates that Needham (1909), Banks (1977), Ghosh (1977, 1980, 1983), Ghosh and Sen (1977), Narasimham (1991), etc. attempted studies related to biodiversity of lacewings from India.

Materials and Methods

Chrysopid adults as well as larvae have been collected from different agroecosystems of Kolhapur region (Study spots) by visiting 15 days interval. The collection was made by one man one hour search method. The collected material was time being placed either in polythene bags or plastic containers perforated by small holes for aeration to the insects collected. The collected material was sorted out at laboratory, examined taxonomically and then identified by consulting literature cited in the references. Adults have been preserved by pinning and larvae by mounting on cardsheet or insect preservative (70 per cent alcohol+glycerine 1:1 proportion). The collection was time being with T. V. Sathe, Dept. of Zoology and deposited in ZSI Kolkata.

Results and Discussion

Results are recorded in Table 9.1. The results indicated that the genera *Chrysoperla* and *Mallada* were most commen and dominant over others in Kolhapur region, specially in sugarcane, horticultural, tobacco, cotton and vegetable ecosystems.

The only general survey for neuropteran fauna in India on *Santhalum album* has been reported by Banks (1933). Ghosh and Sen (1977) reported about dozen of aphidovorous lacewings. They also gave checklist of Indian Planipennia which included by species belonging to 10 genera of the family Chrysopidae. The present work is first attempt on diversity and checklist of lacewings from Kolhapur, Maharashtra. A good number of species are known from the region. There is a strong need for understanding of lacewings diversity and their utility in biological pest control. Therefore, attempts should be made on biological, ecological and ethological aspects of lacewing for better understanding and better utility in IPM strategies of pest control in future.

Acknowledgements

Authors are thankful to Head, Department of Zoology, Shivaji University, Kolhapur for providing facilities to this work.

Table 9.1: Diversity of Lacewings from Agroecosystems of Kolhapur.

Sl.No.	Lacewings	Family	Prey	Occurence	Distrubution
1.	Ankylopteryx octopunctata (Fab.)	Chrysopiae	Aphids	April-Oct.	Karveer, Jaysingpur
2.	Chrysopa infecta Newman	Chrysopiae	Aphids	Jan-April	Karveer, Azara, Shirol, Kagal
3.	Chrysopa bandrina Nivas	Chrysopiae	Aphids	March-April	Kagal, Karveer
4.	Chrysopa barodensis (Navas)	Chrysopiae	Aphids	Jan-March	Karveer, Shirol, Hatkanangale, Gadhinglaj
5.	Chrysopa scelestes Banks	Chrysopiae	Maconellicoccus hirsutus, Bemisia tabaci, Pyrilla purpusilla, Myrus persicae	April-May July.-Oct. Jan-Feb	Karveer, Kagal, Gadhinglaj
6.	Chrysopa amitri Navas	Chrysopiae	Aphids	Jan-April	Karveer, Kagal
7.	Chrysopa villallongae Navas	Chrysopiae	Whiteflies, Aphids, Pyrilla, Aphis craccivora	Aug-Nov	Karveer, Kagal, Shirol
8.	Chrysopa virgestes Banks	Chrysopiae	Aphids, Jassids, whiteflies	Aug-Nov	Gadhinglaj, Karveer, Kagal, Shirol
9.	Chrysopa sp.	Chrysopiae	Aphids, Spodoptera litura	Jan Aug-Nov	Kagal, Karveer, Azara Karveer, Gadhinglaj, Shirol
10.	Chrysopa sp.	Chrysopiae	Thrips tabaci	Aug	Karveer, Kagal
11.	Chrysopa carnea (Stephens)	Chrysopiae	Aphids, whiteflies	July-Nov.	Karveer, Kagal, Azara
12.	Chrysopa punensis	Chrysopiae	Aphids, Thrips	Jan-March	Karveer, Kagal, Azara
13.	Chrysopa sanadensis	Chrysopiae	Helicoverpa armigera	Nov-March	Karveer, Kagal, Ajara, Shirol
14.	Glenochrysopa glorious (Navas)	Chrysopiae	Helicoverpa armigera	Nov-March	Karveer, Kagal, Ajara, Shirol
15.	Mallada bertrani (Navas)	Chrysopiae	Spodoptera sp., Aphids, Whiteflies	July-Nov	Karveer, Kagal, Ajara, Shirol
16.	Mallada herasina (Navas)	Chrysopiae	Aphids	April	Karveer, Kagal, Ajara, Shirol
17.	Mallada khandalensis (Navas)	Chrysopiae	Aphis, Whiteflies, Jassids, Spodoptera, Mealy bugs	July-Jan April-May	Karveer, Radhanagari, Ajara Karveer, Radhanagari

Contd

Table 9.1–Contd...

Sl.No.	Lacewings	Family	Prey	Occurence	Distrubution
18.	*Mallada rocasolanai* (Navas)	Chrysopiae	Potato jassids, Aphids	July–Jan	Karveer, Kagal
19.	*Chrysopa madestes* Banks	Chrysopiae	*Aphis gossypii*	July–Nov	Karveer, Kagal, Radhanagari
20.	*Cintameva cymbelefaciata*	Chrysopiae	Jassids, *Empoasea devastans,*	July–Nov	Karveer, Kagal, Radhanagari
			Whitefly *Bemisia tabaci*	July–Nov	Karveer, Kagal, Radhanagari

References

Banks, N., 1911. Notes on Indian Neuropteroid insects. *Proc. Ent. Soc., Wash.*, 13: 99–106.

Ghosh, S.K., 1977. Fauna of Rajasthan, India–Neuroptera. *Rec. Zool. Surv. India*, 72: 309–313.

Ghosh, S.K., 1980. On a small collection of Neuroptera from Andaman and Nicobar Islands. *Rec. Zool. Surv. India*, 77: 247–254.

Ghosh, S.K., 1983. Neuroptera from Western peninsular India with new records. *Rec. Zool. Surv. India*, 81: 77–87.

Ghosh, S.K. and Sen, S., 1977. Check list of Indian Planinennia (Order: Neuroptera) *Rec. Zool. Surv. India*, 73: 277–326.

Narasimham, A.V., 1991. Biosystematics of Chrysopidae Unpub. *Ann. Rep.* 1990–91. B.C.C., Banglore.

Needham, J.C., 1909. Notes on the Neuropteran in the collection of the Indian Museum. *Res. Ind. Mus.*, 3: 185–210.

Chapter 10

LACEWINGS FOR BIOLOGICAL PEST CONTROL:

(B) Predatory Capacity of *Chrysoperla carnea* (Stephens) (Neuroptera : Chrysopidae) on *Rhopalosiphum maidis* Fitch. Under Laboratory Conditions

T.V. Sathe and S.R. Sawant

*Department of Zoology, Shivaji University,
Kolhapur – 416 004, Maharashtra, India*

ABSTRACT

Chrysoperla carnea (Stephens) (Neuroptera:Chrysopidae) is potential predator of several aphid species. Therefore, predatory potential of *C. carnea* was tested in the laboratory conditions (25±1°C, 70-75 per cent RH, 12 hr photoperiod) against *Rhopalosiphum maidis* Fitch (Hemiptera:Aphididae). *C. carnea* consumed on an average 234 aphids (range 126 to 310). The results indicate that *C. carnea* is very good biocontrol agent of aphid *R. maidis*.

Keywords: Chrysoperla carnea, Predatory potential, Aphid, Rhopalosiphum maidis.

Introduction

Lacewings (Neuroptera : Chrysopidae) are of interest to a large group of entomologists and farmers because of their role as predators of arthropod pests such as insects, spiders and mites in agroecosystems. *Chrysoperla carnea* Stephens (Chrysopidae) feeds on maize aphid *Rhopalosiphum maidis* Fitch in Kolhapur region. Therefore, for hopping biological control of *R. maidis*, predatory potential of *C. carnea* against *R. maidis* has been studied under laboratory conditions. The review of literature indicates that food consumption in Neuroptera has been studied by Steiz and Devetak (1999), Barnes (1975), New (1975), Mibrath *et al.* (1993), Silva *et al.* (2007), etc.

Materials and Methods

Initial culture of Lacewings *C. carnea* was started from collecting adults of *C. carnea* from sugarcane ecosystems of Kolhapur. Field collected adults of *C. carnea* were initially reared on maize aphids *Rhophalosiphum maidis* and allowed to lay eggs on the filter paper placed in petridish (size 38 cm diameter). The newly emerged eggs were separated into either small aerated plastic container or petridish for incubation. The newly emerged larvae were given 2nd instar nymphs of *R. maidis* for the food consumption experiments. All three instars of *C. carnea* given 2nd instar *R. maidis* and food consumption per day was recorded. After 12 hour food was changed. The experiments were conducted under laboratory conditions. (25 $\pm1°C$, 70-75 per cent R.H. and 12 hr photoperiod) and replicated for ten times.

Results

Results are recorded in Table 10.1. The lacewing *C. carnea* completed its life cycle within 30 days. Incubation, larval and pupal durations were 3, 19 and 8 days respectively. Maximum 310 and minimum 126 aphids were consumed by the third instar of *C. carnea*. The minimum and maximum consumption of aphids in I^{st} and IInd instar larvae of *C. carnea* were 23 and 48 and 58 and 170 aphids respectively.

Table 10.1: Predatory Capacity of *C. carnea* against *R. maidis* Under Laboratory Conditions.

Sl.No.	Instar	No. of Aphids Consumed
1.	I	23.00 (Range 22-48)
2.	II	163.00 (Range 58-170)
3.	III	234.00 (Range 126-310)

Discussion

Food quality, quantity and consumption has great importance in mass rearing of any biocontrol agents. Feeding habits of Neuroptera seems to be well understood with the observations of Withycombe (1923) on the biology and several neuropteran species. For years these observations have been cited by many authors and most species have been considered to be rather general feeders, many of them preferred aphids as main prey organisms. Many different food sources such as Jam, honey, pollen, yeast, cooked meat and different soft bodied insects have been used as a food source. Principi (1940,1956), Principi and Canard (1984), Canard *et al.* (1990) investigated feeding habits of adults lacewings.

**Figure 10.1: Predatory Capacity of *C. carnea* against *R. maidis*
Under Laboratory Conditions.**

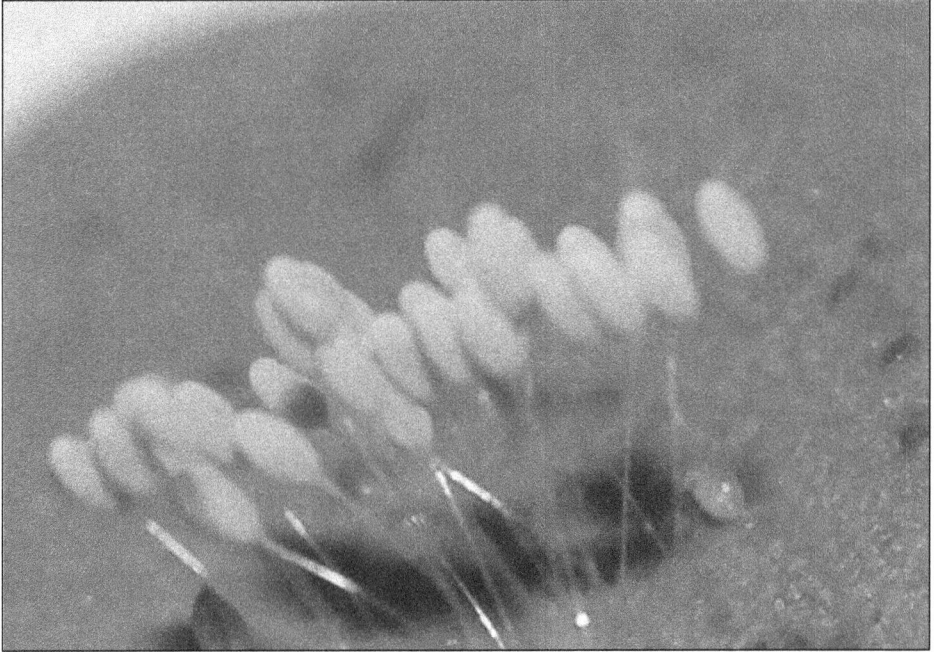

Eggs

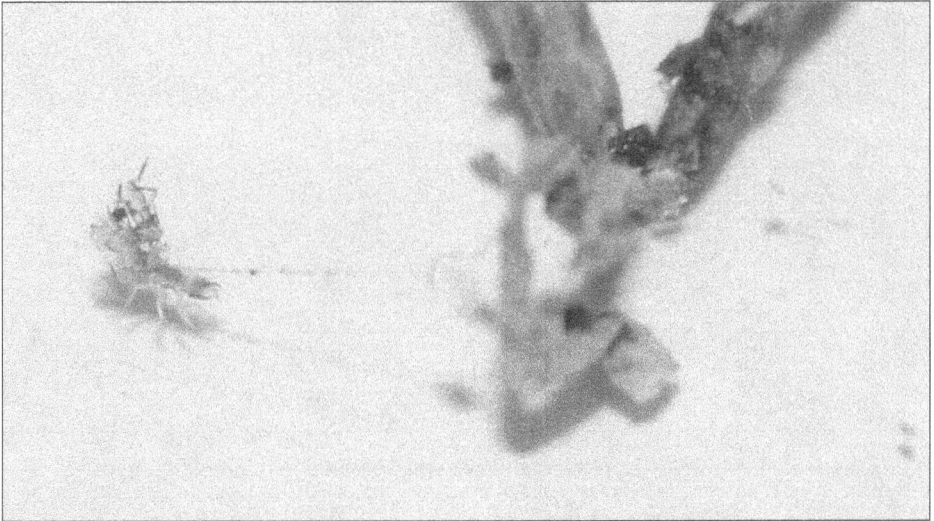

1ˢᵗ instar larva and aphids

Contd...

Figure 10.1–*Contd...*

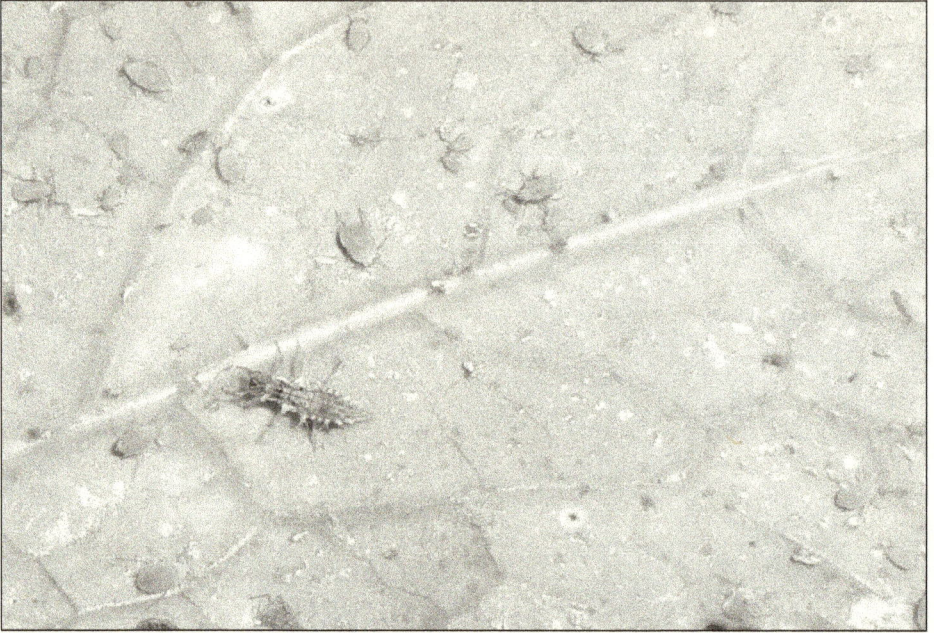

IInd instar larva and aphids

IIIrd instar larva and aphids

In the present study food consumption rate by *C. carnea* against its prey *R. maidis* has been investigated. The results indicated that minimum and maximum consumption was 23 and 163 preys in Ist and IInd instar larvae respectively. IIIrd instar larvae consumed 234 aphids with range of 126 to 310 individuals. The present study will be helpful improving mass rearing technique of *C. carnea*. Upto date several modifications in the rearing techniques of lacewings have been developed (Barnes 1975, Ridgway *et al.*, 1970, Tulisalo 1984, etc.).

Biological control by application of Chrysopidae has been carried out in the field as well as in greenhouses. In the field application many problems, including dispersal of adults, application of methodology, are still unsolved. The present work will add great relevance in that context.

Acknowledgements

Authors are thankful to Prof. and Head, Dept. of Zoology, Shivaji University Kolhapur for providing facilities to this work.

References

Albuaueque, G.S., Tauber, C.A. and Tauber, M.J., 1994. *Chysopa externa* (Neuroptera: Chrysopidae) Life history and potential for biological control in Central and South America. *Biol. Control*, 4: 8–13.

Barnes, B.N., 1975. The life history of *Chryopa zastrowi* Esb–Pet. (Neuroptera: Chrysopidae). *J.Ent. Soc. S. Afr.*, 38: 47–53.

Canard, M., Kokubu, H. and Duelli, P., 1990. Tracheal trunks supplying air to the foregut and feeding habits in adults of Europian green lacewing species (Insecta: Neuroptera: Chrysopidae). In: *Advances in Neuropterology*, (Eds.) M.W. Mansell and H. Aspock. Pretoria, R.S.A., pp. 277–286.

Mibrath, L.M., Tauber, M.J. and Tauber, C.A., 1993. Prey specificity in *Chrysopa*: An interspecific comparison of larval feeding and defensive behavior. *Ecology*, 74: 1384–1393.

New, T.R., 1975. The biology of Chrysopidae and Hemerobiidae (Neuroptera), with reference to their usage as biocontrol agents : review *Trans. R.Entomol. Soc. Lond.*, 127: 115–140.

Principi, M.M., 1940. Contribti allo studio dei 'Neurotteri' italiani I. Chrysopa septempunctata Wesm. E *Chrysopa flavifrons* Brauer. *Boll. Ist Ent*. Univ. Bologna, 12: 63–144.

Principi, M.M., 1956. Contribti allo studio dei 'Neurotteri' italiani XIII. Studio morphologico, etologico e sistematico di un gruppo omogeneo di specie del Gen. *Chrysopa* Leach (*C. flavifrons* Brauer, prasina Burm. E. *clathrataa* Sch.). *Ent. Boll. Ist* Univ. Bologna, 21: 319–410.

Principi, M.M. and Canard, M., 1984. Feeding habits. In: *Biology of Chrysopidae*, (Eds.) M. Canard, Y. Semeria and T.R. New. Dr. W. Junk Publishers, pp. 76–92.

Ridgway, R.L., Morrison, R.K. and Badgley, M., 1970. Mass rearing a green lacewing. *J. Econom. Ent.*, 63: 834–836.

Stelzi, M. and Devetak, D., 1999. Neuroptera in agricultural ecosystems. *Agricut. Ecosyst. and Environ.*, 74: 305–321.

Silva, P.S., Albuquerque, G.S., Tauber, C.A. and Tauber, M.J., 2007. Life history of wide spread Neotropical predator, *Chrysopodes lineafrons* (Neuroptera : Chrysopidae). *Biol. Control*, 41: 33–41.

Tulisalo, U., 1984. Mass rearing techniques. In: *Biology of Chrysopidae*, (Eds.) M. Canard, Y. Semeria and T.R. New. Dr. W. Junk Publishers, pp. 213–220.

Withycombe, C.L., 1923. Notes on the biology of some British Neuroptera (Planipennia). *Trans. Ent. Soc.* London, 1922: 501–594.

Chapter 11

ICHNEUMONIDS IN PEST MANAGEMENT:

(A) Biology of *Diadegma trochanterata* (Morley) (Hymenoptera : Ichneumonidae): An Internal Larval Parasitoid of the Castor Capsule Borer *Dichocrocis punctiferalis* Guene (Lepidoptera : Pyralididae)

T.V. Sathe*

*Department of Zoology, Shivaji University,
Kolhapur – 416 004, Maharashtra, India*

ABSTRACT

Diadegma trochanterata (Morley) (Hymenoptera : Ichneumonidae), an internal larval parasitoid and biocontrol agent of the caster capsule borer *Dichocrocis punctiferalis* Guene (Lepidoptera : Pyralididae). Hence, objecting its utility in biological control of *D. punctiferalis,* biology of *D. trochanterata* has been studied. The parasitoid completed its life cycle from egg to adult within 18.5 days. Adult

* E-mail: profdrtvsathe@rediffmail.com

longevity averaged 15 days and 17 days in males and females respectively with 20 per cent honey. Mating occurred at day time, oviposition takes place within 40-50 seconds. The parasitoid selected 4-5 day old larvae of the host for maximum parasitism, 46.00 per cent under laboratory conditions (25 ±1 °C, 70-75 per cent R.H. and 12 hr photoperiod).

Keywords: Diadegma trochanterata, Parasitoid, Biology, Dichocrocis punctiferalis, Castor pest.

Introduction

Diadegma trochanterata (Morley) (Hymenoptera : Ichneumonidae) is an internal, larval parasitoid of the caster capsule borer *Dicrococis punctiferalis* Guene (Lepidoptera : Pyralididae). Caster *Ricinus communis* L. is oil seed and sericultural crop. The leaves of this crop are used to feed eri silkworms *Samia cynthia ricini*. Hence, control of this pest is essential part from the view of agriculture and sericulture. Chemical control is not without danger to silkworm and secondly, pesticides lead several serious problems like pollution, pest resistance, pest resurgence, secondary pest out break etc. Keeping in view all above facts, present work was carried out. Review of literature indicates that Atwal (1976), Fisher (1959), Gangrade (1964), Tikar and Thakre (1961), Leong and Oatman (1968), Bartell and Pass (1978), Sathe (1988, 1990, 2008) etc. attempted biological studies in Ichneumonid parasitoids of various insect pests.

Materials and Methods

The larvae of *D. punctiferalis* and cocoons of *D. trochantera* have been collected from the fields of caster to maintain laboratory culture of parasitoid. Similarly, host larvae also reared in the laboratory on castor capsule. 4-5 day old 50 larvae of *D. punctiferalis* were exposed to two mated females of parasitoid in glass cage (25 x 25 x 30 cm) for 6 hr. The parasitized larvae were dissected after every 12 hr in normal saline solution and parasitoid eggs and larvae were collected for further observations. The larval stages were treated with 50 per cent chloroform and 50 per cent ethanol and mounted in Hoyer's medium on microslides for biometrical studies. Newly emerged sexes were kept separate in specimen tubes and confined in a pair (1 male and 1 female) into the test tube (10 x 2.5 cm) for mating studies. Similarly, a host larva was taken into test tube and then mated female was released for noting the observations on oviposition. Host age selection was studied by exposing 50 host larvae ranging from 1 day old to 20 day old in glass cage (25 x 25 30 cm) for 12 hr and the parasitoid larvae were then separated and reared on caster capsules under laboratory (25±1 °C, 70-75 R.H.; 12 hr photoperiod). Per cent of parasitism was noted on each day on the host larvae.

Results

Life Cycle

The parasitoid showed five larval instars. First two were caudate and remaining three were hymenopteriform. The life cycle period from egg to adult formation was

18.5 days. The parasitoid showed 4 distinct stages of life cycle *viz*. egg, larva, pupa and adult.

Egg

Eggs were elongately curved and tapered at one end and rounded at another end. The chorion was smoothly opaque and white. Incubation period was 3.5 days. Mostly single egg was laid by the parasitoid on host body but rarely 2 to 3 eggs also laid. In 20 individuals eggs averaged 0.26 mm and 0.062 mm in length and width.

First Instar

First instar larva was creamy white and with long tail *i.e.* caudate type. The tail length was about the half the length of the body. 13 body segments were visible but not clear distinction between thorax and abdomen. The tracheal system was with single longitudinal trunk with several branches. Spiracles were not prominent. The parasitoid consumed internal tissues of the host larva. This stage lasted for 3 days and averaged 1.25 mm in length, 0.16 mm in width and mandibles averaged 0.20 mm in length and 0.014 mm in width.

Second Instar

This stage was opaque with reduced tail and prominent 13 well defined body segments. No vesicle was present. The tracheal system showed two longitudinal trunks with small branches. Spiracles were poorly seen. The larva averaged 1.95 mm in length and 0.36 mm in width. Mandibles in 20 individuals averaged 0.28 mm in length and 0.020 mm in width. The larva was typically curved. This stage lasted for 2 days during which the larva consumed internal tissues of the host larva keeping intact important organs of the host.

Third Instar

Third instar larvae were yellowish opaque in body colour and measured 3.00 mm in length 0.42 mm in width in 20 individuals. Mandibles averaged 0.038 mm in length and 0.027 mm in width. This stage lasted for one day. The tracheal system was with two longitudinal trunks with well developed several branches of tracheae. Spiracles were prominent.

Fourth Instar

This stage was with yellowish opaque coloured and with well defined body segments. Fourth instar was longer than fifth instar and more straight than other instars. No trace of tail was present. The tracheal system was with transverse trunk and two longitudinal trunks with well developed side branches. Spiracles were very prominent. The larvae averaged 4.00 mm in length and 0.73 mm in width and mandibles 0.068 mm in length and 0.034 mm width in 20 individuals. This stage lasted for 2 days and larva consumed host tissues.

Fifth Instar

Fifth instar was also yellowish opaque in colour but was curved and slightly tapered to both the ends and was typically hymenopteriform. The larva consumed internal tissues of the host body and full grown in about 2.5 days. Tracheal system was more developed with more prominent spiracles. The larvae averaged 3.6 mm in length and 0.76 mm in width and mandibles averaged 0.070 mm in length 0.055 mm in width in 20 individuals.

Cocoon

Cocoon was dirty white coloured, rounded at both the ends and averaged 4.2 mm in length and 1.10 mm in width. The parasitoid emerged from anterior side of the cocoon by taking circular cut to the cocoons.

Pupa

Pupa was exarate type, whitish in body colour at initial stage but became dark brown approaching its full development. Pupae averaged 3.8 mm and 0.76 mm in length and width respectively in 20 individuals. Pupal duration averaged 6 days.

Adults

Adult males were smaller than females and females with ovipositor for egg laying. The ovipositor was curved upward and was not straight. Head, thorax and some part of dorsal side of abdomen was black in colour.

Adult Longevity

The males and females survived for an average of 15 days and 17 days respectively with 20 per cent honey. In control, the parasitoid died within 2 days and with water the parasitoid extended their survival for 3 days.

Mating

Mating occurred at day time, about afternoon. Mating period averaged 2.45 min.

Oviposition

Parasitoid oviposited 4.5 days old host larvae immediately when confined in test tubes. Oviposition was completed within 35 seconds.

Host Age Selection

The parasitoid selected 4-5 day old host larvae for maximum progeny production (46.00 per cent). The hosts 1-2 day old and 15 to 20 day old were rejected by the parasitoid for parasitism. The progeny production was possible from hosts of 3 day old to 12 day old hosts. The sex ratio was favouring females.

Discussion

Fisher (1959) reported 4 instars in *Horogenes chrysostictes* (Gemelin) (Ichneumonidae), a parasitoid of *Ephestia sericarium* (Scott.). Similarly, Tikar and Takre (1961) reported 4 instars in *H. fenestralis* Holmgren, a parasitoid of common caterpillar. While, Gangrade (1964) and Sathe (2008) reported five instars in *Campoletis chlorideae* Uchida, an internal larval parasitoid of *Helicoverpa* (=*Heliothis*) *armigera* (Hubn.). In an Ichneumonid parasitoid *Diadegma argenteopilosa* (Cameron), a parasitoid of

Spodoptera litura (Fab.) also reported 5 instars. The present findings are in agreement with number of instars reported by Gangrade (1964) and Sathe (2008) in respective parasitoid species. The life cycle from egg to adult was completed within 17 and 18 days in *D. trichoptilus* (Cameron) and *D. argenteopilosa* (Cameron) respectively (Sathe 1988; Sathe, 2008) while in the present form *D. trochanterata* the life cycle was completed within 20 days. *D. trochanterata* mate and oviposit immediately in the laboratory and selects 4-5 day old caterpillars for maximum parasitization. The above attributes of *D. trochanterata* will add great relevance in designing mass rearing technique for the same parasitoid in future.

Acknowledgement

Author is thankful to U.G.C., New Delhi for providing financial assistance to the UGC Major Resarch Project F. No. 37-334/2009 (SR).

References

Atwal, A.S., 1976. *Agricultural Pests of India and South East Asia*. Kalyani Publ., New Delhi, p. 303.

Bartell, D.P. and Pass, B.C., 1978. Morphology, development and behaviour of the immature stages of the parasite *Bathyplectes curculionis*. *Ann. Ent. Soc. Am.*, 7(1): 23–30.

Fisher, R.C., 1959. Life history and ecology of *Horogenes chrysostictes* (Gemelin) (Hymenoptera : Ichneumonidae), a parasitoid of *Ephestia sericarium* (Scott.) (Lepidoptera : phycilidae). *Can. J.*, 37: 429–446.

Gangrade, G.A., 1964. On the biology of *Campoletis perdistinctus* Uchida (Hym; Ichneumonidae) in Madhya Pradesh. *Ann. Ent. Soc. Am.*, p. 231–244.

Leong, G.K.L. and Oatman, E.R., 1968. The biology of *Camplex haywardi* (Hym.; Ichneumonidae), a primary parasitoid of the potato tuber worm. *Ann. Ent. Soc. Am.*, 6: 26–36.

Sathe, T.V., 1988. Biology of *Diadegma trichoptilus* (Cameron) (Hymenoptera : Ichneumonidae), a larval parasitoid of *Exelastis atomosa* Walsingham. *J. Curr. Biosci.*, 5(2): 37–40.

Sathe, T.V., 1990. The Biology of *Diadegma argenteopilosa* (Cameron) (Hymenoptera : Ichneumonidae), an internal larval parasitoid of *Spodoptera litura* (Fab.). *The Entomologist*, 109(1): 2–7.

Sathe, T.V., 2008. Mass production technique for *Camoletis chlorideae* Uchida. *Biotechnological Approaches in Entomology*, 3: 64–74.

Sathe, T.V. and Jadhav, A.D., 2001. *Sericulture and Pest Management*. Daya Publishing House, New Delhi, pp. 1–197.

Tikar, D.T. and Thakre, K.R., 1961. Bionomics and biology of immature stages of an Ichneumonid *Horogenes fenestralis* Holmgren, a parasitoid of common caterpillar. *Indian J. Ent.*, 23: 116–124.

Chapter 12

ICHNEUMONIDS IN PEST MANAGEMENT:

(B) Effect of *Dichocrocis punctiferalis* Guenee (Lepidoptera : Pyralididae) Density on Progeny Production of *Diadegma trochanterata* (Hymenoptera : Ichneumonidae)

T.V. Sathe

*Department of Zoology, Shivaji University,
Kolhapur – 416 004, Maharashtra, India*

ABSTRACT

Dichocrocis punctiferalis Guenee (Lepidoptera : Pyralididae) is Castor capsule borer and difficult to control with conventional pesticides. An Ichneumonid parasitoid *Diadegma trochanterata* (Morley) is larval parasitoid of *D. punctiferalis* and acts as good biocontrol agent. Therefore, optimum host density was investigated for maximum progeny production. 10, 25, 50, 75 and 100 host densities have been tried, out of which 50 host density was proved to be optimum for maximum progeny production, 56.80 per cent.

Keywords: D. trochanterata, Progeny production, Biological control, Host, D. punctiferalis.

Introduction

Diadegma trochanterata (Morley) (Hymenoptera : Ichneumonidae) is an internal, larval parasitoid of the Castor capsule borer *Dichocrocis punctiferalis* Guenee (Lepidoptera : Pyralididae). Castor *Ricinus communis* L. is very good oil seed and sericultural crop of India. Eri silkworms are reared on castor leaves. Therefore, ecofriendly control of caster plant is an important aspect of pest management. The parasitoid attacked 10-13 per cent *D. punctiferalis* larvae from the field of Western Maharashtra (Sathe, 1992). Host density selection by parasitoid plays a very crucial role in mass rearing of biocontrol agents and further field release for pest control. Keeping in view all above facts, present work was carried out. Perusal of literature indicates that parasitoid host age has been studied by Yeargan and Latheef (1976), Leong and Oatman (1968), Oatman and Platner (1974), Sathe (1992), Sathe and Nikam (1985), Sathe and Margaj (2001), etc.

Materials and Methods

Initial cultures of pest and parasitoid were maintained in the laboratory (25±1°C, 70-75 per cent, 12 hr. photoperiod) by collecting cocoons of parasitoids and larvae of *D. punctiferalis* from fields. 10, 25, 50, 75 and 100 host densities have been tried. The larvae of each density were exposed to mated parasitoid in glass cage, (30 x 25 x 25 cm) for 6 hr. and then parasitized larvae were transferred in plastic containers and reared separately. The hosts and parasitoids were fed with castor capsules and 20 per cent honey respectively during the conduct of experiments. Each experiment was replicated for five times.

Results

Results are recorded in Table 12.1. For maximum progeny production (56.80 per cent) 50 host density was proved to be the best. The host densities10, 25, 75 and 100 yielded 40.00 per cent, 35.50 per cent, 48.00 per cent, and 47.00 per cent, progeny production respectively under laboratory conditions (25±1°C, 70-75 per cent, RH 12 hr. photoperiod). The host densities 75 and 100 also showed good potential for progeny production.

Discussion

According to Leong and Oatman (1968) in *Campoplex haywardi* Blanchard (Ichneumonidae), larval parasitoid of *Phthorimaea operculella* (Zeller), the host density was 75 larvae per tuber. Their observations indicate that there was great variation between replicates at the same larval densities and they further observed that mean emergence at higher densities was from 3 per cent to 15 per cent less than at 75 host density. In the present study there was not much variation in the progeny production in replicates of each host density, probably due to the unfluctuating laboratory conditions.

Oatman and Platner (1974) studied density dependant relationship between an Ichneumonid *Temeluchu* sp. and a pest tuber moth *P. operculella*. They exposed hosts in 50, 100 and 150 densities and found 150 host density as optimum for maximum progeny production.

Table 12.1: Host Density Influence on Progeny Production in *D. trochanterata*.

Sl.No.	Host Density	Number of Replicates	Numbers of Individuals Obtained			Per cent Moth Emergence	Per cent Parasitoids Emergence
			Moths	Parasitoids	Both		
1.	10	5	20 (10)	20	40	40.00	40.00
2.	25	5	50 (16)	44	94	40.00	35.50
3.	50	5	99 (9)	142	241	39.6	56.80
4.	75	5	180 (15)	180	360	48.00	48.00
5.	100	5	250 (15)	235	285	50.00	47.00

Figures in parenthesis denotes mortality number.

Yeargan and Latheef (1976) also studied host parasitoid density relationship in a model *Hypera postica* (Gyllenhall) (Coleoptera : Curculionidae) and *Bathyplectes curculionis* (Thompson) (Hymenoptera : Ichneumonidae). Their study revealed that the parasitoid behavioural response to uneven host densities was only one component of its overall responses to host abundance which does not rule out the possibilities of a dependent relationship between *H. postica* and *B. curculionis*. Our findings are also in agreement with Yeargan and Latheef (1976).

Sathe (1985) exposed early instar larvae of *Exelastis atomosa* Wals. to *D. trichoptilus* in densities of 10, 20, 30, 40 and 50 for 24 hr. He recorded highest number of parasitoids emerged with host density 30 while, in the present form the optimum host density was 50 which yielded highest, 56.80 per cent progeny production. The present work will be helpful for mass rearing of *D. trochanterata* as baseline data in future.

Acknowledgement

Author is thankful to U.G.C., New Delhi for providing financial assistance to the Major Research Project F. No. 37-334/2009 (SR).

References

Leong, G.K.L. and Oatman, E.R., 1968. The biology of *Campoplex haywardi* (Hymenoptera : Ichneumonidae), a primary parasite of the potato tuber worm. *Ann. Ent. Soc. Am.*, 61(1): 26–36.

Oatman, E.R. and Platner, G.R., 1974. The biology of *Temelucha* sp., *Platensis* group (Hymenoptera : Ichneumonidae), a primary parasite of the potato tuber worm. *Ann. Ent. Soc. Am.*, 67: 275–280.

Sathe, T.V., 1992. Natural enemies of some insect pests of economic importance. *Oikoassay*, 9: 15–17.

Sathe, T.V. and Margaj, G.S., 2001. *Cotton Pests and Biocontrol Agents*. Daya Publishing House, Delhi, pp. 1–166.

Sathe, T.V. and Nikam, P.K., 1985. Influence of host density on percentage parasitism by *Diadegma trichoptilus* (Cameron), a larval parasitoid of *Exelastis atomosa* (Wals.). *Indian J. Parasitol*, 9(2): 229–230.

Yeargan, K.V. and Latheef, M.A., 1976. Host parasitoid density relationship between *Hypera postica* (Coleoptera : Curculionidae) and *Bathyplectes curculionis* (Hymenoptera : Ichneumonidae). *J. Kanasas Cant.*, 49: 551–556.

Chapter 13

ICHNEUMONIDS IN PEST MANAGEMENT:

(C) Host Specificity in *Pristomerus testaceus* Morley (Ichneumonidae : Hymenoptera)

T.V. Sathe*

*Department of Zoology, Shivaji University,
Kolhapur – 416 004, Maharashtra, India*

ABSTRACT

Pristomerus testaceus Morley (Ichneumonidae : Hymenoptera) is larval parasitoid of *Euzophera perticella* Rag. and *Leucinodus orbonalis* Guenee in Kolhapur region. Therefore host specificity has been studied by providing *E. perticella, L. orbonalis, Sylepta derogata* and *Eutectona machearalis*. The order of preference for hosts given by the parasitoid was *L. orbonalis > E. perticella > Sylepta derogata > E. machearalis*. The results showed that *P. testaceus* is good biocontrol agent in agro and forest ecosystem.

Keywords: *Pristomerus testaceus*, Host specificity, Biocontrol agent.

* E-mail: profdrtvsathe@rediffmail.com

Introduction

Host specificity in the parasitic Hymenoptera discloses many of the factors that determine whether or not any two given species are to be associated as host and parasitoid, while there are a few species, that limit their attack to a single host species. Most parasitoids are found in nature to attack several different hosts (Sathe and Margaj, 2001).

Parasitoids may bread in the laboratory on unnatural hosts is often of great importance in the mass culture of parasitoids or field colonization. Certain insects are more amenable to insectary production than others and when these can be used as factitious hosts the mass culture of the parasitic species is simplified and can often be put on a commercial basis.

Parasitoid limits its attack to the suitable host species that occur in nature, has stimulated investigation. The species of host preferred by a parasitoid may vary depending on the host species and the climatic conditions (Weshloh, 1981). The odour of a host may result for acceptability of host. In some cases, the novel hosts are suitably accepted. Some non host species are readily attacked when encountered. The studies regarding a parasitoid preference for different host species have been reported by Gutierrez (1970), Lingren *et al.* (1970), Lingren and Noble (1972), Drooz and Fedde (1972), Calvert (1973), Jackson *et al.* (1979), Hopper and King (1984) etc. in parasitic hymenoptera.

The objective of the present work was to search for new hosts and determine the degree of susceptibility of some of the more important hosts of this parasitoid.

Materials and Methods

To find out host preference given by the parasitoid species, the laboratory reared four day old caterpillars of *Euzophera perticella* (Rag.), *Leucinodus orbonalis* (Guenee), *Sylepta derogata* and *Eutectona machearalis* were exposed in glass cages (with mixed host density), 50 to single mated female of *Pristomerus testaceus*. After 24 hr exposure, the larvae were separated in plastic containers and reared for parasitoid/moth emergence. The experiments were replicated for 5 times for confirming the results. During the course of study the parasitoids were fed with 50 per cent honey and hosts with their respective food plant parts. In all cases mortality (unknown) was less than 10 per cent. The experiments were conducted at laboratory conditions (25±1°C, 66-70 per cent R.H., 12 hr photoperiod).

Results

The larvae of *Euzophera perticella* and *Leucinodus orbonalis* were accepted by parasitoid *P. testaceus* for parasitization. The order of preference for hosts given by the parasitoid was *L. orbonalis* > *E. perticella* > *Sylepta derogate* > *E.machearalis*. The results (Table 13.1) showed that *P. testaceus* is good biocontrol agent for above pest in agro and forest ecosystem.

Table 13.1: Host Specificity in *Pristomerus testaceous*.

Sl.No.	Host Species	Total Hosts Exposed	Parasitoids Progeny			Per cent Parasitism
			Male	Female	Total	
1.	Leucinodusorbonalis	250	32	50	82	32.80
2.	Euzopheraperticella	250	35	55	90	36.00
3.	Sylepta derogate	250	21	24	45	18.00
4.	Eutectonamachearilis	250	23	27	50	20.00

Discussion

The theoretical approaches *i.e.*, switching and optimal selection are found in the literature which are applied to the host selection by parasitoids. The tendency of parasitoids to concentrate on the most abundant host species has been called switching. Where switching will occur depending on the preference of a parasitoid or one of the host species. Preference is usually measured as a deviation in the proportion of host attack and the host available in the environment. The optimal host selection models predict that when a parasitoid has choice between two host species, it should always accept the best on in terms of fitness gain and that the second should only be accepted if it would increase number of offspring produced per time unit. The less profitable species should either be accepted or rejected. Lingren *et al.* (1970) reported 27 species of Lepidoptera as the hosts of *Campoletis perdistinctus* Morley. Unfortunately, none of the hosts were as well adapted to mass rearing. However, the alalfa Looper, ball worm, cabbage looper, fall armyworm, southern army worm, tobacco bud worm and variegated cutworms were very susceptible to parasitism by. *C. perdistinctus*. Fall armyworm was also adoptable for mass rearing of the parasitoid than the other hosts. Lingren and Noble (1972) also studied the host preference towards the larvae of noctuids *viz.* beet army worm. *Spodoptera exigua* (Hubner), ball worm *Heliothis zea* (Boddie), Cabbage looper, *Trichoplusia ni* (Hubner), fall army worm *S. frugiperda* (Smith) southern army worm, *Prodenia eridania* (Cramer), tobacco bud worm *H. virescens* (F.) and yellow stripped army worm, *Prodenia ornithogalli* Guenee. They found that the boll worms, fall army worms and tobacco bud worms were mostly preferred in all laboratory tests, whereas, the cabbage looper was the least and no parasitoid cocoons were produced from the larvae of the beet army worms. Their laboratory and field tests indicated that, the boll worms and tobacco bud worms were among the most preferred noctuid hosts. In the present study, among the exposed four species of hosts, *P. testaceus* showed preference in the order *L. orbonalis* > *E. perticella* > *Sylepta derogata* > *E. machearalis*.

The parasitoid should have a wide host range in order to increase its potentiality in the biological control programme. The present data will be helpful for mass culture of the parasitoid species and for field colonization.

Acknowledgement

The author are thankful to Head, Department of Zoology, Shivaji University, Kolhapur for providing the facilities and U.G.C., New Delhi for providing financial assistance to the Major Research Project F. No. 37-334/2009 (SR).

References

Calvert, D., 1973. Experimental host presence of *Minoctonus paulensis* including a hypothetical scheme of host selection. *Ann. Ent. Soc. Am.*, 66: 28–33.

Drooz, A.T. and Feddle, H.V., 1972. Discriminate host selection by *Monodontomerus dentipes*. *Environ. Entomol.*, 1: 522–23.

Gutierrez, A.P., 1970. Studies on host selection and host specificity of the aphid hyperparasite, *Charips victrix* S. Host selection. *Ann. Ent. Soc. Am.*, 63: 1495–1498.

Hopper, K.R. and King, E.G., 1984a. Preference of *Microplitis croceipes* (Hym : Braconidae) for instars and species of *Heliothis* spp. (Lepidoptera : Noctuidae). *Environ. Entomol.*, 15: 1145–1150.

Hopper, K.R. and King, E.G., 1984 b. Feeding and movement on cotton of *Heliothis* spp. (Lepidoptera : Noctuidae). Parasitized by *Microplitis croceipes* (Hym : Braconidae). *Environ. Entomol.*, 13: 1654–1660.

Iwasam, Y. Suziki and Matsuda, 1984. Theory of oviposition strategy of parasitoids I effect of mortality and limited egg number. *Theoretical Population Biology*, 26: 205–227.

Jackson, C.G., Neemann, E.G. and Patana, R., 1979. Parasitization of 6 lepidopteran cotton pests by *Chilonus blackburnii* (Hym : Braconidae). *Entomophaga*, 24(1): 99–105.

Lewis, W.J. and Vinson, S.B., 1971. Sutability of certain *Heliothis* spp. (Lepidoptera : Noctuidae) hosts for the parasite, *Cardiochiles nigriceps*. *Ann. Ent. Soc. Am.*, 64: 970–972.

Lingren, P.D. and Noble, L.W., 1972. Preference of *Compoletis perdistinctus* for certain *Heliothis* spp. Lepidoptera : Noctuid larvae. *J. Econ. Entomol.*, 65: 104–107.

Salt, G., 1935. Experimental studies in insect parasitism–III Host selection. *Proc. Roy. Soc. Ser. B. Biel. Sci.*, 117: 413–435.

Tawfik, M.F.S., 1957. Host parasite specificity in a braconid, *Apanteles glomeratus* L. *Nature*, 179: 1031–1032.

Thorpe, W.H. and Jones, F.G.W., 1937. Olffactory conditioning in a parasitic insect and its relation to the problem of host selection. *Proc. Roy. Entomol. Soc.* London Ser B., 124: 56–81.

Vinson, S.B., 1975. Biochemical coevolution between parasitoids and their hosts. In: *Evolutionary Strategies of Parasitic Insects and Mites*, (Ed.) p. 14–48, P.W. Drica, New York, Plenum, 1–225 pp.

Weshloh, R.M., 1981. Host location by parasitoids in *"Semichemicals their Role in Pest Control"* (Eds.) D.A. Nordhund, R.L., Jones and W.J. Lewis, pp. 79–95.

Chapter 14

ICHNEUMONIDS IN PEST MANAGEMENT:

(D) Life Cycle of *Netelia ephippiata* Smith (Ichneumonidae: Hymenoptera) on Castor Semilooper *Achea janata* (Linnaeus)

T.V. Sathe* and N. Nilam Shendage

*Department of Zoology, Shivaji University,
Kolhapur – 416 004, Maharashtra, India*

ABSTRACT

Netelia ephippiata Smith (Ichneumonidae: Hymenoptera) is an internal, larval parasitoid of caster semilooper *Achea janata* (Linnaeus) (Noctuidae : Lepidoptera). The parasitoid completed its life cycle within 19.5 days. Egg, larval and pupal periods were 3.5, 9 and 7 days respectively.

Keywords: Netelia ephippiata, Parasitoid, Biocontrol agent, Achea janata.

* E-mail: profdrtvsathe@rediffmail.com

Introduction

Netelia ephippiata (Hymenoptera : Ichneumonidae) is an internal, larval parasitoid of the castor semilooper *Achea janata* (Lepidoptera : Noctuidae). Castor *Ricinus communis* L. is oil seed and sericultural crop. The leaves of this crop are used to feed eri silkworms *Samia cynthia ricini*. Hence, control of this pest is essential part from the view of agriculture and sericulture. Chemical control is not without danger to silkworm and secondly pesticides lead several serious problems like pollution, pest resistance, pest resurgence, secondary pest out break etc. Keeping in view all above facts, present work was carried out. Review of literature indicates that Atwal (1976), Fisher (1959), Gangrade (1964), Tikar and Thakre (1961), Leong and Oatman (1968), Bartell and Pass (1978), Sathe (1988, 1990, 2008) etc. attempted biological studies in Ichneumonid parasitoids of various insect pests.

Materials and Methods

The larvae of *A. janata* and cocoons of *N. ephippiata* have been collected from the fields of castor to maintain laboratory culture of parasitoid. Similarly, host larvae also reared in the laboratory on caster leaves. 3-4 day old 50 larvae of *A. janata* were exposed to two mated females of parasitoid in glass cage (25 x 25 x 30 cm) for 6 hr. The parasitized larvae were dissected after every 12 hr in normal saline solution and parasitoid eggs and larvae were collected for further observations. The larval stages were treated with 50 per cent chloroform and 50 per cent ethanol and mounted in Hoyer's medium on microslides for biometrical studies. Newly emerged sexes were kept separate in specimen tubes and confined in a pair (1 male and 1 female) into the test tube (10 x 2.5 cm) for mating studies. Similarly / a host larva was taken into test tube and then mated female was released for noting the observations on oviposition. Parasitized hosts were reared in plastic containers for life cycle studies. Daily a lot of 20 parasitized larvae were dissected for confirming incubation / larval instars and pupal periods.

Results

Life Cycle (Figures 14.1 A–F)

The parasitoid showed five larval instars. First two were candate and remaining three were hymenopteriform. The life cycle period from egg to adult formation was 19.5 days. The parasitoid showed 4 distinct stages of life cycle *viz.* egg, larva, pupa and adult.

Egg (Figure 14.1A)

Eggs were elongately curved and tapered at one end and rounded at another end. The chorion was smoothly upaque and white. Incubation period was 3.5 days. Mostly single egg was laid by the parasitoid on host body but rarely 2 to 3 eggs also laid. In 20 individuals eggs averaged 0.22 mm and 0.060 mm in length and width.

First Instar (Figure 14.1B)

First instar larva was creamy white and with long tail *i.e.* caudate type. The tail length was not about the half the length of the body. 13 body segments were visible

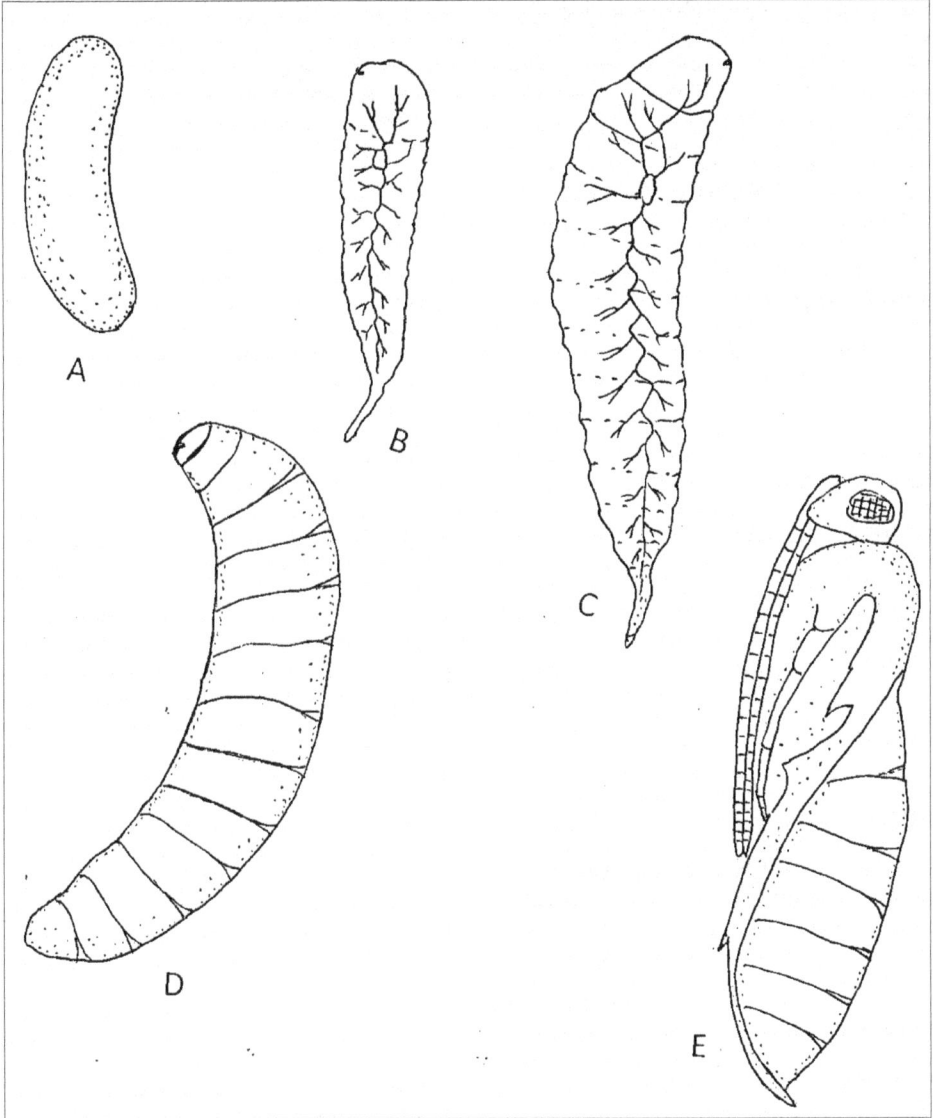

Figure 14.1: *Netelia ephippiata*
A: Egg; B: First instar; C: Second instar; D: Fifth instar; E: Pupa.

but not clear distinction between thorax and abdomen. The tracheal system was with single longitudinal trunk with several branches. Spiracles were not prominent. The parasitoid consumed internal tissues of the host larva. This stage lasted for 3 days and averaged 2.23 mm in length, 0.14 mm in width and mandibles averaged 0.20 mm in length and 0.014 mm in width.

Second Instar (Figure 14.1C)

This stage was opaque with reduced tail and prominent 13 well defined body segments. No vesicle was present. The tracheal system showed two longitudinal trunks with small branches. Spiracles were poorly seen. The larva averaged 1.90 mm in length and 0.32 mm in width. Mandibles in 20 individuals averaged 0.26 mm in length and 0.020 mm in width. The larva was typically curved. This stage lasted for 2 days during which the larva consumed internal tissues of the host keeping intact important organs of the host.

Third Instar

Third instar larvae were yellowish opaque in body colour and measured 3.20 mm in length 0.40 mm in width in 20 individuals. Mandibles averaged 0.035 mm in length and 0.025 mm in width. This stage lasted for one day. The tracheal system was with two longitudinal trunks and with well developed several branches of tracheae. Spiracles were prominent.

Fourth Instar

This stage was with yellowish opaque coloured and with well defined body segments. Fourth instar was longer than fifth instar and more straight than other instars. No trace of tail was present. The tracheal system was with transverse trunk and two longitudinal trunks with well developed side branches. Spiracles were very prominent. The larvae averaged 4.10 mm in length and 0.71 mm in width and mandibles 0.067 mm in length 0.033 mm width in 20 individuals. This stage lasted for 2.5 days and larva consumed host tissues.

Fifth Instar (Figure 14.1D)

Fifth instar was also yellowish opaque in colour but was curved and slightly tapered to both the ends and was typically hymenopteriform. The larva consumed internal tissues of the host body and full grown in about 3 days. Tracheal system was more developed with more prominent spiracles. The larvae averaged 3.8 mm in length and 0.75 mm in width and mandibles averaged 0.070 mm in length and 0.052 mm in width in 20 individuals.

Cocoon

Cocoon "was dirty white coloured, rounded at both the ends and averaged 4.5 mm in length and 1.25 mm in width. The parasitoid emerged from anterior side of the cocoon by taking circular cut to the cocoons.

Pupa (Figure 14.1E)

Pupa was exarate type, whitish in body colour at initial stage but became dark brown approaching its full development. Pupae averaged 3.9 mm and 0.77 mm in length and width respectively in 20 individuals. Pupal duration averaged 6 days.

Adults (Figure 14.1F)

Adult males were smaller than females and females with ovipositor for egg laying.

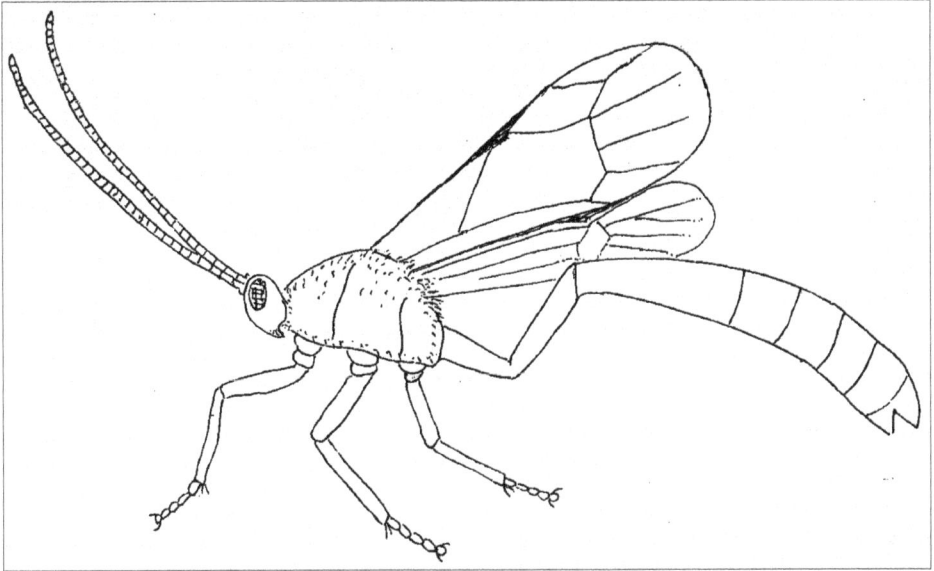

Figure 14.1: *Netelia ephippiata*
F: Female.

Adult Longevity

The males and females survived for an average of 17 days and 19 days respectively with 20 per cent honey. In control/the parasitoid died within 2 days and with water the parasitoid extended their survival for 3 days.

Mating

Mating occurred at day time/about afternoon. Mating period averaged 2.40 min.

Oviposition

Parasitoid oviposited 3-4 day old host larvae immediately when confined in test tubes. Oviposition was completed within 35 seconds.

Host Age Selection

The parasitoid selected 3-4 day old host larvae for maximum progeny production.

Discussion

Fisher (1959) reported 4 instars in *Horogenes chrysostictes* (Gemelm) (Ichneumonidae), a parasitoid of *Ephestia sericarium* (Scoit.). Similarly, Tikar and Takre (1961) reported 4 instars in *H. fenestralis* Holmgren, a parasitoid of common caterpillar. While, Gangrade (1964) and Sathe (2008) reported five instars in *Campoletis chlorideae* Uchida, an internal larval parasitoid of *Helicoverpa* (= *Heliothis*) *armigera* (Hubn.). In an Ichneumonid parasitoid *Diadegma argenteopilosa* (Cameiron), a parasitoid of *Spodoptera litura* (Fab.) also reported 5 instars. The present findings are in agreement with number of instars reported by Gangrade (1964) and Sathe (2008) in respective parasitoid species. The life cycle from egg to adult was completed within 17 and 18 days in D. *trichoptilus* (Cameron) and *D. argenteopilosa* (Cameron) respectively (Sathe 1988; Sathe, 2008) while in the present form *N. ephippiata* the life cycle was completed within 19.5 days. *N. ephippiata* mate and oviposit immediately in the laboratory and selects 3-4 day old caterpillars for maximum parasitization. The above attributes of *N. ephippiata* will add great relevance in designing mass rearing technique for the same parasitoid in future.

Acknowledgement

Author is thankful U.G.C., New Delhi for providing financial assistance to the UGC Major Research Project F. No. 37-334/2009 (SR).

References

Atwal, A.S., 1976. *Agricultural Pests of India and South East Asia*. Kalyani Publ., New Delhi, p. 303.

Bartell, D.P. and Pass, B.C., 1978. Morphology, development and behaviour of the immature stages of the parasite *Bathyplectes curculionis. Ann. Ent. Soc. Am.,* 7(1): 23–30.

Fisher, R.C., 1959. Life history and ecology of *Horogene schrysostictes* (Gemelin) (Hymenoptera : Ichneumonidae), a parasitoid of *Ephestia sericarium* (Scott.) (Lepidoptera : Phycilidae). *Can. J.,* 2001; 37: 429–446.

Gangrade, G.A., 1964. On the biology of *Campoletis perdistinctus* Uchida (Hym; Ichneumonidae) in Madhya Pradesh. *Ann. Ent. Soc. Am.,* pp. 231–244.

Leong, G.K.L. and Oatman, E.R., 1968. The biology of *Camplex haywardi* (Hym.; Ichneumonidae), a primary parasitoid of the potato tuber worm. *Ann. Ent. Soc. Am.,* 6: 26–36.

Sathe, T.V., 1988. Biology of *Diadegma trichoptilus* (Cameron) (Hymenoptera : Ichneumonidae), a larval parasitoid of *Exelastis atomos* Walsingham. *J. Curr. Biosci.,* 5(2): 37–40.

Sathe, T.V., 1990. The Biology of *Diadegma argenteopilosa* (Cameron) (Hymenoptera : Ichneumonidae), an internal larval parasitoid of *Spodoptera litura* (Fab.). *The Entomologist,* 109(1): 2–7.

Sathe, T.V., 2008. Mass production technique for *Campoletis chlorideae* Uchida. *Biotechnological Approaches in Entomology,* 3: 64–74.

Sathe, T.V. and Jadhav, A.D., 2001. *Sericulture and Pest Management*. Daya Publishing House, New Delhi, pp. 1–197.

Tikar, D.T. and Thakre, K.R., 1961. Bionomics and biology of immature stages of an Ichnneumonid *Horogenes fenestralis* Holmgren, a parasitoid of common caterpillar *Indian J. Ent.,* 23: 116–124.

Chapter 15

ICHNEUMONIDS IN PEST MANAGEMENT:

(E) Life Cycle of *Xanthopimpla pedator* (Hymenoptera : Ichneumonidae): A Parasitoid of Tasar Silkworm *Antheraea mylitta* (Lepidoptera : Saturnidae)

T.V. Sathe, N.N. Shendage, D.K. Jadav,
A.M. Bhosale and Sheetal Londhe

Department of Zoology, Shivaji University,
Kolhapur – 416 004, Maharashtra, India

ABSTRACT

Xanthopimpla pedator (Hymenoptera : Ichneumonidae) is an internal larval-pupal parasitoid of Tasar silkworm *Antheraea mylitta* (Lepidoptera: Saturnidae). The parasitoid parasitized last instar silkworms of *A.mylitta* and fifth instar of parasitoid emerged from cocoons (pupae) of the silkworms. The parasitoid completed its life cycle within 18 days at laboratory conditions (25±1°C, 70-75 per cent R.H; 12 hr photoperiod).

Keywords: Xanthopimpla pedator, Parasitoid, Silkworm Antheraea mylitta, Life cycle.

Introduction

Xanthopimpla pedator (Hymenoptera: Ichneumonidae) is internal larval-pupal parasitoid of Tasar silkworm *Antheraea mylitta* (Lepidoptera: Saturnidae). The parasitoid is uesful biocontrol agent of Jowar, Paddy and Sugarcane stem borers *Chilo* spp. (Figure 15.1F) However, it also parasitizes tasar silkworms and acts as pest of sericultural crop. The parasitoid has been previously reported on *A. mylitta* by Sathe and Jadhav (2001). At present, it is not so destructive pest of silkworms, but may attain the full status of the pest on this very economically important silkworm. Life cycle studies in Ichneumonid parasitoids have been attempted by Fisher (1959), Gangrade (1964), Tikar and Thakre (1961), Sathe (1988, 1990, 2008), etc.

Materials and Methods

Cocoons of *X. pedator* were collected from Paddy, Maize and Sugarcane fields and also from silkworm rearing houses from Kolhapur region and initial culture of parasitoid was started by providing 5th instar (last instar) larvae *Antheraea mylitta* to parasitoid. Newly emerged males and females were kept in separate test tubes and then a pair (1 male and 1 female) was confined into the test tube. Immediately mateing occurred in the test tube. Mated females were exposed to 25 larvae of *A. mylitta* for oviposition for 6hr and then worms were separared and reared by providing Ain (*Terminalia tomentosa*) leaves to silkworms. Thus, sufficient number of parasitized larvae were collected and then dissected for immature stages regularly after 12hr interval and instars have been identified on the basis of increase in mandible size and head capsule. The parasitoid larva came out of the silk worm by breaking the body or by cutting the cocoons of the silkworm. Incubation period was counted from ovipositon to egg hatching, Larval period was counted from formation of first instar to transformation of pupa. Pupal period was counted from the formation of pupa to adult emergence. The parasitoid was given 20 per cent honey solution and to *A. mylitta* larvae Ain leaves as food during the conduct of experiments.

Results

Egg

The ovarian eggs were some what elongately curved and tapering from one end and broder from another. The chorion was smoothly opaque and white. In 20 individuals eggs averaged 0.70mm and 0.15mm in length and width respectively. Generally only one egg was deposited by the female but occasionally 2 or 3 may be placed. Incubation period averaged 3.5 days.

Larvae

Five instars have been recorded in *X. pedator*. First two were caudate type and last three were hymenopteriform. In 20 individuals average larval period was 8.5 days.

First Instar (Figure 15.1A)

The larva was creamy white or opaque white in body colouration. 13 post cephalic segments have been observed. The larva showed characteristic long tapering

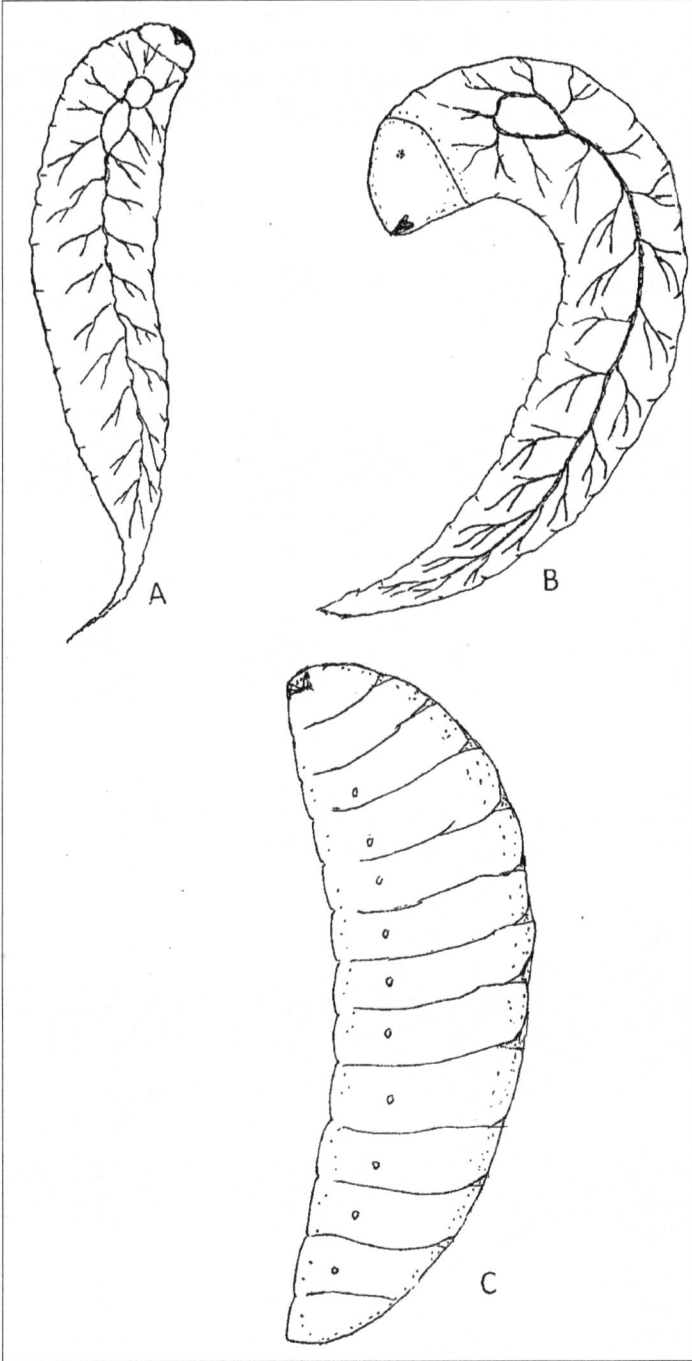

Figure 15.1: *Xanthopimpla pedator*
A: First instar; B: Second instar; C: Fifth instar.

tail. The head side was slightly tapered but posterior was sufficiently tapered. In 20 individuals mandible averaged 0.050mm and 0.35mm in length and width and head capsule 0.40mm and o.48mm in length and width respectively. The larval length and width averaged 2.5mm and 0.38mm respectively. Tracheal system comprised two longitudinal trunks.

This stage lasted for 2days.

Figure 15.1D: *Xanthopimpla* **(Female).**

Figure 15.1E: *Xanthopimpla* **(Male).**

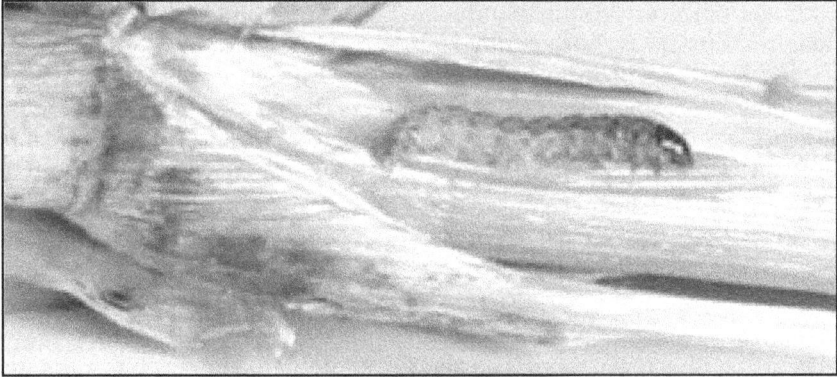

Figure 15.1F: *Chillo partellus.*

Second Instar (Figure 15.1B)

It was opaque with 13 cephalic segments. Tail was present but some what reduced. However, in late stage body size enlarged considerably. Average length and width in 20 individuals were 5.0mm and 1.10mm respectively. The mandible averaged 0.075mm and 0.55mm and head capsule averaged 0.60mm and 0.56mm in length of width respectively. This stage lasted for 2 days. Tracheal system was with two longitudinal trunks and with more branched. Head and tail region was reduced in size. The larva consumed internal tissues of the silkworm.

Third Instar

The body segments of this stage were clearly seen, this stage was opaque white in colour. No trace of tail was noticed in this stage. However, both ends of the larva were tapered. Posterior was more tapered and curved. The tracheal system was well developed with transverse trunk connected to longitudinal one and was more branched. The larval body averaged 5.5mm and 1.21mm in length and width respectively. While, mandibles and head capsule averaged 0.082mm and 0.065mm and 1.12mm and 1.00mm in length and width respectively. This stage stayed for 1day.

Fourth Instar

The body segments of this stage were very distinct, no tail was present and the larva was opaque whitish in colour. This stage lasted for 1.5 days. Fourth instar was relatively straight and longer than fifth instar stage. The larva was more as parallel sided at middle and slightly tapered at both the ends. However, both ends of larva were rounded. Tracheal system was well developed and with more tracheal branches. Spiracles were very clear in this stage. The average length and width in 20 individuals were 12mm and 2.15mm respectively and mandibles averaged 0.12mm and 0.90mm in length and width respectively.

Fifth Instar (Figure 15.1C)

This stage was typically hymenopteriform curved, and tapered towards both the ends and with very strong mandibles. The body colour was yellowish opaque. The length of larva was shortend than fourth stage but width was found increased. This stage lasted for 2.5 days and consumed internal tissues of the silkworm considerably by keeping organ systems intact. The larvae averaged 10.5mm and 2.80mm in length and width in 20 individuals and 0.20mm and 1.10mm in mandibles respectively. The head capsule averaged 2.35mm and 1.80mm in length and width respectively. The fifth instar came out of the body of silkworm by breaking body wall and spun dirty cocoon.When late last instars were parasitized 5th instar of parasitoid came from cocoon of the silkworm.

Cocoon

The cocoons were barrel shaped and dirty coloured. The cocoons measured for 14mm and 3.5mm in length and width respectively.

Pupa

Initially pupa was whitish in appearence with all three distinct parts *viz.* head, thorax and abdomen. However, older froms became darker and darker finally became dark brown to yellow. The adult take longitudinal cut to the cocoon at anterior side and emerged from the cocoon, spread wings, cleaned the body, make the typical movements of legs, abdomen, antennae and ovipositor, and performed fluttering of wings and flew away within 30 to 60 minutes after emergence.

Adults (Figures 15.1D and 15.1E)

Males were without ovipositor and smaller than females and females showed long straight ovipositor for egg laying. Both sexes were yellowish with black spots on their thorax and abdomen.

Discussion

In *Horogenes chrysostictes* (Gemelin)(Ichneumonidae), a parasitoid of *Ephestia sericarium* (Scott.) (Lepidoptera: Phycitidae). Fisher (1959) reported five instars. Tikar and Thakare (1961) reported only four instars and later, Gangrade (1964) and Sathe (2008) reported five instars in the *C. chlorideae*, a parasitoid of *H. armigera*. Sathe (1990) also recorded five instars in an Ichneumonid parasitoid *Diadegma argenteopilosa* Cameron, an internal, larval, solitary parasitoid of *Spodoptera litura* Fab. (Lepidoptera: Noctuidae). In the present Ichneumonid form, *X. pedator* also five larval instars were noticed. The life cycle from egg to adult was completed in 17 days and 18 days in *Diadegma trichoptilus*, (Cameron) and *D. argenteopilosa* (Sathe,1988, 2008.) respectively. While *X. pedator* completed its life cycle within 18 days under laboratory conditions (25±1°C,70-75 per cent R.H. and 12hr photoperiod). The present work will be useful for protecting silkworms from this parasitoid and controlling *Chilo* species in agro ecosystems by providing basic information on life cycle of the parasitoid.

Acknowledgement

Authors are thankful to Prof. and Head Dr. T.V. Sathe for providing facilities to this work and U.G.C. New Delhi for providing financial assistance to the Major Research Project- F.No.37-334/2009(SR).

References

Fisher, R.C., 1959. Life history and ecology of *Horogenes chrysostictes* (Gemelin) (Hymenoptera: Ichneumonidae), a parasitoid of *Ephestia sericarium* (Scott.) (Lepidoptera: phycilidae).*Can. J. Zool.*, 37: 429–446.

Gangrade, G.A., 1964. On the biology of *Campoletis perdistinctus* Uchida (Hym; Ichneumonidae) in Madhya Pradesh. *Ann. Ent. Soc. Am.*, 33: 231–244.

Sathe, T.V., 1988. Biology of *Diadegma trichoptilus* (Cameron) (Hymenoptera: Ichneumonidae), a larval parasitoid of *Exelastis atomosa* Walsingham. *J. Curr. Biosci.*, 5(2): 37–40.

Sathe, T.V., 1990. The Biology of *Diadegma argenteopilosa* (Cameron) (Hymenoptera: Ichneumonidae), an internal larval parasitoid of *Spodoptera litura* (Fab.). *The Entomologist*, 109(1): 2–7.

Sathe, T.V., 2008. Mass production technique for *Campoletis chlorideae* Uchida. *Biotechnological Approaches in Entomology*, 3: 64–74.

Sathe, T.V. and Jadhav, A.D., 2001. *Sericulture and Pest Management.* Daya Publishing House, New Delhi, pp. 1–197.

Tikar, D.T. and Thakare, K.R., 1961. Bionomics and biology of immature stages of an Ichneumonid *Horogenes fenestralis* Holmgren, a parasitoid of common caterpillar. *Indian J. Ent.*, 23: 116–124.

Chapter 16

BRACONIDS IN PEST MANAGEMENT:

(A) Life Cycle of *Apanteles euproctisiphagus* Muzaffar (Hymenoptera : Braconidae): A Parasitoid of Castor Hairy Caterpillar *Euproctis lunata* (Walker)

P.M. Bhoje[1], T.V. Sathe[2] and Nilam Shendge[2]

*[1]Department of Zoology, Y.C. Mahavidhyalaya,
Warnanagar, Kolhapur, Maharashtra, India
[2]Department of Zoology, Shivaji University,
Kolhapur – 416 004, Maharashtra, India*

ABSTRACT

Apanteles euproctisiphagus Muzaffar (Hymenoptera : Braconidae), is internal larval parastoid of castor hairy caterpillar *Euproctis lunata* (Walker). The parastoid completed its life cycle from egg to adult within 15 days, incubation period 3 days, larval 7 days and pupal period 5 days.

Keywords: *Apanteles euproctisiphagus, Parasitoid, Life cycle, Euproctis lunata.*

Introduction

Apanteles euproctisiphagus Muzaffar (Hymenoptera: Braconidae) is internal, larval parasitoid of *Euproctis lunata* (Walker) (Lepidoptera: Noctuidae) and acts as good biocontrol agent. Biological control is very good alternative for chemical since chemical control leads several serious problems such as pollution, killing of beneficial organisms, pest resistance, pest resurgence, secondary pest outbreak, interruption to ecocycles, etc. Therefore, hoping the biological control of *E. lunata*, the present work was undertaken. This work will remain as basic data for mass rearing of *A. euproctisiphagus* in biological control programme. In past, Boodryk(1969), Cardona and Oatman (1971), Rojas- Rousse and Benoit (1977), Sathe and Nikam (1985), Sathe (1987a&b), etc. attempted life cycle studies in braconid parasitoids.

Materials and Methods

Initial culture of parasitoid was started from collecting parasitized larvae of *E. lunata* from caster field. The parasitoids were screened and mated females (2) exposed to 4day old 50 host larvae for 2hr in a glass cage, 30×25×25cm. Parasitized larvae were separated for further development and observations. The parasitoid eggs and larvae were collected after 6 hr interval dissecting parasitized larvae in normal saline. Instars were identified by increased size of mandibles and spiracles and life cycle data has been traced.

Results

Egg

Hymenopteriform and thin walled eggs were white and deposited into the host body from dorsal side. 2-3eggs were randomly deposited in the body of single larva. Incubation occurred within 3 days after oviposition. The eggs averaged 0.160mm in length and 0.48mm in width. The eggs were tapering from one side and broader from another side. Tapering end was introduced into the host body through ovipositor by the female.

Instars

The parasitoid showed three larval instars. First two were vesiculate and third was hymenopteriform.

Ist Instar

First instar was with broad quadrate head, and with 3 thoracic and 7 abdominal segments. First instar was characterized by vesicle at the posterior end which was with single layered epithelial cells. This stage lasted for 2.5 days. One armed tracheal system was also noticed in this stage. Mandibles in 20 individuals averaged 0.039 mm and 0.017 mm in length and width respectively.

IInd Instar

Second instar was cylindrical, straight, with head, thorax and abdomen indistinctly cleared but with 3 thoracic and 7 abdominal clear segments. These whitish opaque coloured second instars lasted for 2 days and found feeding on internal tissues of host larvae. Caudal vesicle was present in this stage. In 20 individuals, mandibles average 0.068 mm and 0.025 mm in body length and width respectively and head capsule averaged 0.21 mm and 0.19 mm in length and width respectively.

IIIrd Instar

This instar lasted for 2.5 days. Mandibles in 20 individuals averaged 0.082 mm and 0.035mm in length and width respectively while head capsule arranged 0.35 mm and 0.30mm respectively. Third instar was opaque coloured and was typically hymenopteriform. The larva tapered slightly towards both the ends. No vesicle was present at the posterior end. The full grown larva emerged from the host body by killing it. The newly emerged parasitoid larva immediately started constructing cocoon around itself. A silvery white cocoon was spun by the third instar larva.

Cocoon

Cocoon was cylindrical and rounded at both the ends and was with silken threads. Parasitoid take circular cut at the posterior end of cocoon for adult emergence and escaped from the cocoon. The circular open end and circular cap was found remained attached with the cocoon body after parasitoid emergence. In 20 individuals cocoon averaged 3.6 mm and 1.4 mm in length and width respectively.

Pupa

Pupa was exarate type; it was light yellow initially except for the blackish eyes and brown ocelli. The pupa became blackish brown approaching development of adult. This stage lasted for 5 days. In 20 individuals pupae averaged 3.4 mm and 1.3 mm in body length and width respectively.

Adult Longevity

With 20 per cent honey solution parasitoid survided for 10 days. In control (without food) the parasitoid died within 1 to 2 days.

Discussion

Atwal (1976) reported nine hymenopterous parasitoids parasitizing *A. janata*, out of which six were braconids. In *Cotesia orientalis* C. and N. (Braconidae), a parasitoid of tur plume caterpillar *Exelastis atomosa* Walsingham (Pterophoridae) Sathe (1987) reported 3 instars. The first two were vesiculate and third was hymenopteriform. Same situation was noticed in *A. euproctisiphagus*. Immediately prior to hatching, the mature embryo straightened out its body and with the help of mandibles it ruptured. As the instars progressed the length and width of mandibles and head capsule also increased relatively but did not showed any co-relations with the age of the larval body form; but larval body length and width showed significant (p<0.01) co-relation with larval age and larval body forms. Same results are obtained in the present form and the findings of Sathe (1987) in parasitoid *C. orientalis*, are in agreement with present form. Rojas-Rousse and Benoit (1977) and Sathe and Nikam (1985) also confirmed the same principle in *Pimpla instigator* (F.) and *Cotesia flavipes* (Cameron) respectively.

In an Ichneumonid parasitoid *Campoletis chlorideae* Uchida the larval instars were five, first two were caudate type and rest were hymenopteriform. In the present braconid parasitoid only 3 instars were present and first two were vesiculate and last (third) was hymenopteriform. Pupae in all above parasitoids were exarate type as noticed in present form. In an Ichneumonid *Diadegma trichoptilus* (Cameron), a parasitoid of *E. atomosa* the entire larval body was converted into cocoon and no larval emergence outside host body was noticed while, in the present study, parasitoid broken the body wall of host and the spun cocoon outside the host body (Sathe,1987).

The present work concludes that the present data on the parasitoid will be helpful for developing mass rearing technique for *A. euproctisiphagus*.

Acknowledgements

Authors are thankful to UGC, New Delhi for financial assistance to Major Research Project F-37-334/2009 (SR)/37-1/2009 (MS) (SR) and ShivajiUniversity for providing facilities.

References

Atwal, A.S., 1976. *Agricultural Pests of India and South-East Asia*. Kalyani Publ., New Delhi, p. 303.

Boodryk, S. B., 1969. The biology of *Chelonus* (*Micro chelonus*) *curvi maculatus* Cameron (Hymenoptera: Braconidae). *J. Ent. Soc. South Africa*, 32: 169–189.

Cardona, C. and Oatman, E.R., 1971. Biology of *Apantles dingus* (Hymenoptera: Braconidae), a primary parasite of the Potato pinworm. *Ann. Ent. Soc. Am.*, 60: 996–1007.

Rojas-Rousse and Benoit, M., 1977. Morphology and biometry of larval instars of *Pimpla instigator* (F.) (Hymenoptera: Ichneumonidae). *Bull. Ent. Res.*, 67: 129–141.

Sathe, T.V. and Nikam, P.K., 1985. Morphology and biometry of immature stages of *Cotesia flavipes*(Cameron) (Hymenoptera: Braconidae), an internal, larval parasitoid of *Chilo partellus* (Swin.). *Indian J. Zool.*, 13(2): 43–46.

Sathe, T.V., 1987a. Morphology and biometry of immature stages of *Cotesia orientalis* C. and N. (Hymenoptera: Braconidae), an internal, larval parasitoid of *Exelastis atomosa* Walsingham. *Uttar Pradesh J. Zool.*, 7: 200–203.

Sathe, T.V., 1987b. Morphology and biometry of immature stages of *Diadegma trichoptilus*(Cameron) (Hymenoptera: Ichneumonidae), an internal, larval parasitoid of *Exelastis atomosa* Walsingham. *Indian J. Zool.*, 15: 29–32.

Chapter 17

BRACONIDS IN PEST MANAGEMENT:

(B) Life Cycle of *Apanteles sudanus* Wilkinson (Hymenoptera : Braconidae): An Internal Larval Parasitoid of Castor Semilooper *Achea janata* Linnaeus (Lepidoptera : Noctuidae)

T.V. Sathe and Nilam Shendge

Department of Zoology, Shivaji University,
Kolhapur – 416 004, Maharashtra, India

ABSTRACT

Apanteles sudanus Wilkinson (Hymenoptera: Braconidae) is an internal larval parasitoid of Castor semilooper *Achea janata* Linnaeus (Lepidoptera: Noctuidae) and acts as biocontrol agent of above pest. Therefore, life cycle of *A. sudanus* was studied in the laboratory conditions (25±1°C, 70-75 per cent R.H., 12hr photoperiod). The parasitoid completed its life cycle within 16 days. Incubation, larval and pupal durations were 3 days, 8 days and 5 days respectively. The adult parasitoids were given 20 per cent honey solution and the *A. janata* castor leaves as food during the conduct of experiments.

Keywords:*Apanteles sudanus, Parasitoid, Life cycle, Achea janata, Pest.*

Introduction

Apanteles sudanus Wilkinson (Hymenoptera: Braconidae) is internal, larval parasitoid of *Achea janata* (Lepidoptera: Noctuidae) and acts as good biocontrol agent of *A.janata*. Biological control is very good alternative for chemical since chemical control leads several serious problems. Therefore, hoping the biological control of *A. janata*, the present work was undertaken. This work will remain as basic data for mass rearing of *A. sudanus* in biological control programme. In past, Boodryk (1969), Cardon and Oatman (1971), Rojas- Rousse and Benoit (1977), Sathe and Nikam (1985), Sathe (1987 a&b), etc. attempted life cycle studied in braconid parasitoids (Figures 17.6 and 17.11).

Materials and Methods

Initial culture of parasitoid was started from collecting parasitized larvae of *A. janata* (Figure 17.5) from castor field. The parasitoids were screened and mated females (2) exposed to 4day old 50 host larvae for 2hr in a glass cage, 30×25×25cm. Parasitized larvae were separated for further development and observations. The parasitoid eggs and larvae were collected after 6 hr interval dissecting parasitized larvae in normal saline. Instars were identified by increased size of mandibles and spiracles and life cycle data has been traced.

Results

Egg (Figure 17.1)

Hymenopteriform and thin walled eggs were white and deposited into the host body from dorsal side. 2-3eggs were randomly deposited in the body of single larva. Incubation occurred within 3 days after oviposition. The eggs averaged 0.160mm in length and 0.48mm in width. The eggs were tapering from one side and broader from another side. Tapering end was introduced into the host body through ovipositor by the female.

Instars

The parasitoid showed three larval instars. First two were vesiculate and third was hymenopteriform.

Ist Instar

First instar was with broad quadrate head, and with 3 thoracic and 7 abdominal segments. First instar was characterized by vesicle at the posterior end which was with single layered epithelial cells. This stage lasted for 3 days.One armed tracheal system was also noticed in this stage. Mandibles in 20 individuals averaged 0.038mm and 0.016mm in length and width respectively.

IInd Instar (Figure 17.2)

Second instar was cylindrical, straight, with head, thorax and abdomen indistinctly cleared but with 3 thoracic and 7 abdominal clear segments. These whitish opaque coloured second instars lasted for 2 days and found feeding on internal

Figures 17.1–17.4: *Apanteles sudanus.*

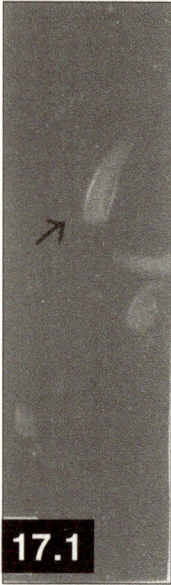

Figure 17.1:   Figure 17.2: Second Instar Larva.   Figure 17.3: Pupa
   Egg.

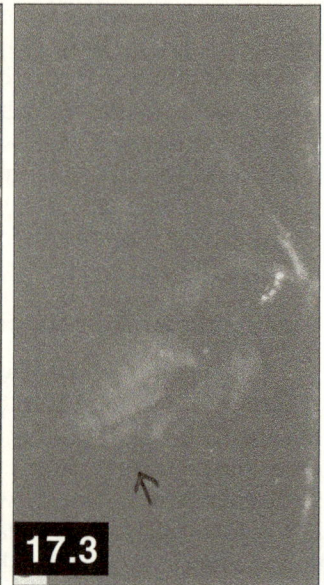

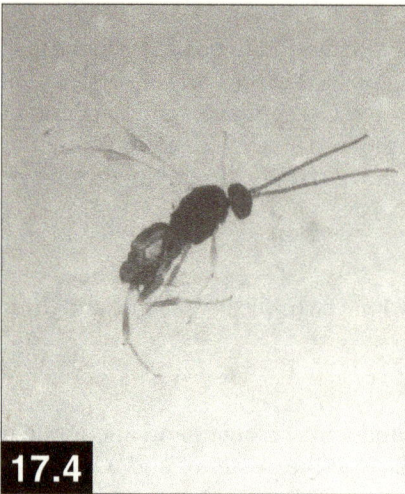

Figure 17.4: Adult Female.   Figure 17.5: *A. janata.*

tissues of host larvae. Caudal vesicle was present in this stage. In 20 individuals, mandibles averageed 0.065 mm and 0.022 mm in body length and width respectively and head capsule averaged 0.20 mm and 0.18 mm in length and width respectively.

Figure 17.6–17.11: Brachonids in Insect Pest Control.

Figure 17.7: Moth (*M. separata*).

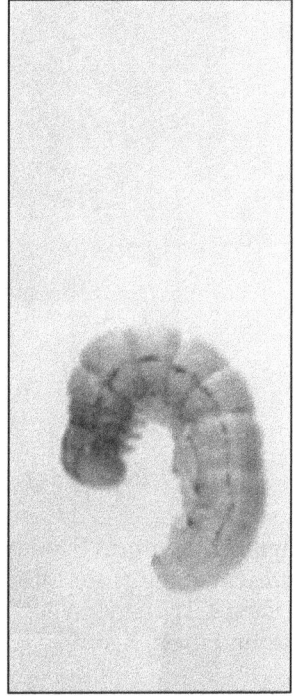

Figure 17.7: Larva (*M. separata*).

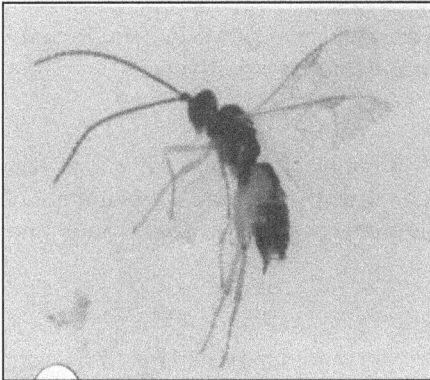

Figure 17.8: *Dolicogenidea*.

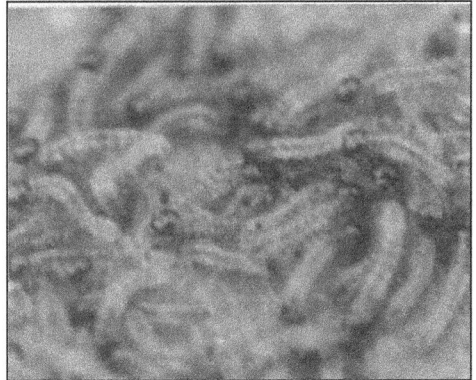

Figure 17.9: Larvae (*S. litura*).

IIIrd Instar

This instar lasted for 3 days. Mandibles in 20 individuals averaged 0.080mm and 0.035mm in length and width respectively while head capsule arranged 0.34mm and 0.30mm respectively. Third instar was opaque coloured and was typically

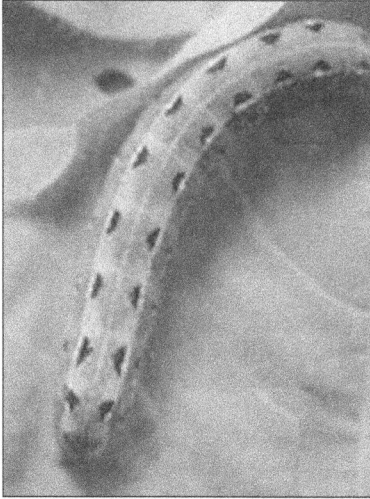

Figure 17.10: Larva (*S. litura*) Fifth Instar

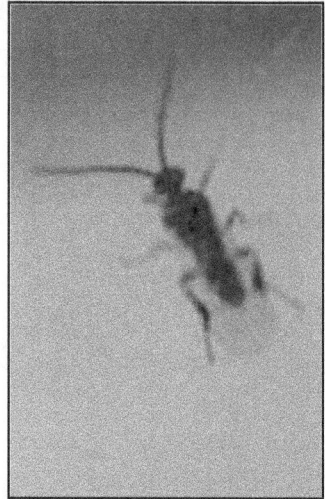

Figure 17.11: *Apanteles prodeniae*

hymenopteriform. The larva tapered slightly towards both the ends. No vesicle was present at the posterior end. The full grown larva emerged from the host body by killing it. The newly emerged parasitoid larva immediately started constructing cocoon around itself. A silvery white cocoon was spun by the third instar larva.

Cocoon

Cocoon was cylindrical and rounded at both the ends and was with silken threads. Parasitoid take circular cut at the anterior end of cocoon for adult emergence and escaped from the cocoon. The circular open end and circular cap was found remained attached with the cocoon body after parasitoid emergence. In 20 individuals cocoon averaged 3.5 mm and 1.3 mm in length and width respectively.

Pupa (Figure 17.3)

Pupa was exarate type; it was light yellow initially except for the blackish eyes and brown ocelli. The pupa became blackish brown approaching development of adult. This stage lasted for 5 days. In 20 individuals pupae averaged 3.3mm and 1.2mm in body length and width respectively.

Adult (Figure 17.4)

Longevity with 20 per cent honey solution parasitoid survided for 10 days. In control (without food) the parasitoid died within 1 to 2 days.

Discussion

Atwal (1976) reported nine hymenopterous parasitoids parasitizing *A. janata*, out of which six were braconids. However, he has not reported the biology of *A. sudanus*. In *Cotesia orientalis* C. and N. (Braconidae), a parasitoid of tur plume caterpillar *Exelastis atomosa* Walsingham (Pterophoridae) Sathe (1987) reported 3 instars. The first two were vesiculate and third was hymenopteriform. Same situation was noticed in *A. sudanus*. As the instars progressed the length and width of mandibles and head capsule also increased relatively but did not showed any co-relations with the age of the larval. This indicates that highly chitinized parts increased in size only after moulting. Same results are obtained in the present form and the findings of Sathe (1987) in parasitoid *C.orientalis* are in agreement with present form. Rojas-Rousse and Benoit (1977) and Sathe and Nikam (1985) also confirmed the same principle in *Pimpla instigator* (F.) and *Cotesia flavipes* (Cameron) respectively.

In an Ichneumonid parasitoid *Campoletis chlorideae* Uchida the larval instars were five, first two were caudate type and rest were hymenopteriform. In the present braconid parasitoid *A. sudanus* only 3 instars were present and first two were vesiculate and last (third) was hymenopteriform. Pupae in all above parasitoids were exarate type as noticed in present form. In an Ichneumonid *Diadegma trichoptilus* (Cameron), a parasitoid of *E.atomosa* the entire larval body was converted into cocoon and no larval emergence outside host body was noticed while, in the present study, parasitoid broken the body of host and the spun individual cocoon outside the body of host (Sathe,1987). The present work concludes that the present data on the parasitoid will be helpful for developing mass rearing technique for *A.sudanus*.

Acknowledgement

Authors are thankful to UGC, New Delhi for financial assistance to Major Research Project F-37-334/2009 (SR)/37-1/2009 (MS) (SR) and Shivaji University for providing facilities.

References

Atwal, A.S., 1976. *Agricultural Pests of India and South-East Asia*. Kalyani Publ., New Delhi, p. 303.

Boodryk, S.B., 1969. The biology of *Chelonus* (*Microchelonus*) *curvimaculatus* Cameron (Hymenoptera: Braconidae). *J. Ent. Soc. South Africa,* 32: 169–189.

Cardona, C. and Oatman, E.R., 1971. Biology of *Apantles dingus* (Hymenoptera: Braconidae), a primary parasite of the Potato pinworm. *Ann. Ent. Soc. Am.,* 60: 996–1007.

Rojas-Rousse and Benoit, M., 1977. Morphology and biometry of larval instars of *Pimpla instigator* (F.) (Hymenoptera: Ichneumonidae). *Bull. Ent. Res.,* 67: 129–141.

Sathe, T.V. and Nikam, P.K., 1985. Morphology and biometry of immature stages of *Cotesia flavipes* (Cameron) (Hymenoptera: Braconidae), an internal, larval parasitoid of *Chilo partellus* (Swin.). *Indian J. Zool.,* 13(2): 43–46.

Sathe, T.V., 1987a. Morphology and biometry of immature stages of *Cotesia orientalis* C. and N. (Hymenoptera: Braconidae), an internal, larval parasitoid of *Exelastis atomosa* Walsingham. *Uttar Pradesh J. Zool.,* 7: 200–203.

Sathe, T.V., 1987b. Morphology and biometry of immature stages of *Diadegma trichoptilus* (Cameron) (Hymenoptera: Ichneumonidae), an internal, larval parasitoid of *Exelastis atomosa* Walsingham. *Indian J. Zool.,* 15: 29–32.

Chapter 18

TACHINIDS IN PEST MANAGEMENT:

(A) Diversity of Tachinids from Agroecosystems of Kolhapur District

T.V. Sathe[1], P.M. Bhoje[2], A.S. Desai[1] and Nilam Shendge[1]

Department of Zoology, Shivaji University,
Kolhapur – 416 004, Maharashtra, India
[2]Department of Zoology, Y.C. Warana College,
Warananagar, Maharashtra, India

ABSTRACT

Tachinids are potential biocontrol agents of several agricultural insect pests. About 10,000 species are described under the family Tachinidae from different corners of the world.However, very little attention is given on this useful group of insects. Hence, biodiversity of Tachinid flies attacking agricultural crops has been studied from Kolhapur district. In all, **26** species of genera *Sturmia, Exorista, Tachina, Drino, Blepharipa,* etc. have been reported parasitizing the pests such as *Spodoptera litura, Spodoptera exigua, Helicoverpa armigera, Achea janata, Anomis flava, Ypsilon flammatra, Acherontia styx* etc. and silkworms *Bombyx mori* and *Antheraea mylitta.*

Keywords: Diversity, Tachinids, Parasitoids pests, Kolhapur district.

Introduction

Tachinids (Diptera : Tachinidae) are the largest and one of the most diverse group of insects found mostly in terrestrial environments. Adults are often brightly coloured with yellow, black orange-red markings or vivid metallic green, blue or other tints and measuring from 2mm to 20mm in body length.Tachinid larvae are parasitoids of several lepidopterous and coleopterons insect pests found in various agro ecosystems thus, acts as biocontrol agents of insect pests (Sathe, 2004). About 10,000 species of tachinids have been reported from different parts of the world (O'Hara, 2008). From oriental region 725 species have been reported (Crosskey, 1976; O'Hara, 2008) and from India 216 species have been recorded belonging to 113 genera (O'Hara, 2009; Lahiri, 2003, 2006; Sathe 2012).

Results and Discussion

The results are recorded in Table 18..1.

The tachinid flies parasitized larval stages of various insect pests such as *Helicoverpa armigera* Hubn, *Spodoptera litura, Achea janata, Ypsilon flammatra,*

Figure 18.1: *Blepharipa* sp.

Figure 18.2: *Dexia* **spp.**

Mythimna separata, Acherontia styx, Anomis flava, Spodoptera exigua and silkworms like *Bombyx mori* L. and *Antheraea mylitta* Drury during the course of study (2010-2012).

In all **26** species of tachinid flies have been reported from various agro ecosystems of Kolhapur. *E. bombycis* parasitized *A. janata, S. litura, H. armigera, Samia Cynthia ricini* Boised in India (Narayanswami *et al.*, 1993), found in most of the months in year and mostly associated with silkworms. However, abundant genera of the district refer to *Exorista, Blepharipa, Drino, Sturmiopsis* and *Tachin*. Tachinid flies are not well

Figure 18.3: *Drino* sp.

Figure 18.4: *Exorista* sp.

Table 18.1

Sl.No.	Species	Tribe	Sub-family	Family	Hosts	Occurrence
1.	*Dexia fulvifera* Von Roder, 1893	Dexini	Dexiinae	Tachinidae	Unknown Lepidopteran larvae	May-June
2.	*Prosena facialis* Curran, 1929	Dexini	Dexiinae	Tachinidae	Unknown Lepidopteran larvae	May-June
3.	*Exorista deligata* Pandelle, 1896	Exoristini	Exoristinae	Tachinidae	Unknown Lepidopteran larvae	May-October
4.	*Exorista sorbillans* (Wiedemann, 1830)	Exoristini	Exoristinae	Tachinidae	*Bombyx mori* *Helicoverpa armigera*	May-October
5.	*Exorista bombycis* (Louis)	Exoristini	Exoristinae	Tachinidae	*Bombyx mori* *Helicoverpa armigera*	May-October
6.	*Blepharipa zebina* Walker (1849)	Goniinii	Exoristinae	Tachinidae	*Antheraea mylitta*	May-January
7.	*Blepharipa zebina manipurensis* Lahiri, 2006	Goniinii	Exoristinae	Tachinidae	*Antherarea mylitta*	May-January
8.	*Blepharipa albocinta* Mensil, 1956	Goniinii	Exoristinae	Tachinidae	*Antheraea mylitta*	May-January
9.	*Exorista castanea* (Wulp, 1894)	Exoristini	Exoristinae	Tachinidae	*Antheraea* spp	May-June
10.	*Exorista japonica* (Townsend, 1909)	Exoristini	Exoristinae	Tachinidae	*Bombyx mori*	Throughout year
11.	*Exorista larvarum* (Linnaeus, 1849)	Exoristini	Exoristinae	Tachinidae	Lepidopteran Caterpillar	August-November
12.	*Dexia prakritae* (Lahiri, 2006)	Dexini	Dexiinae	Tachinidae	Lepidopteran Caterpillar	August-November
13.	*Dexia quadristriata* (Lahiri, 2006)	Dexini	Dexiinae	Tachinidae	Lepidopteran Caterpillar	August-November
14.	*Dexia divergens* (Walker, 1856)	Dexini	Dexininae	Tachinidae	*Acherontia styx*	August-November
15.	*Carcelia caudata* (Baranov, 1931)	Eryciini	Exoristinae	Tachinidae	*Agrotis ypsilon*	August-November
16.	*Drino argenticeps* (Macquart, 1851)	Eryciini	Exoristinae	Tachinidae	*Spodoptera litura*	August-November
17.	*Drino facialis* (Townsend, 1928)	Eryciini	Exoristinae	Tachinidae	*Spodoptera litura*	August-November
18.	*Drino solennis* (Walker, 1858)	Eryciini	Exoristinae	Tachinidae	*Spodoptera exigua*	May-October
19.	*Sturmiopsis inferens* (Townsend, 1916)	Eryciini	Exoristinae	Tachinidae	*Chillo* spp	May-October
20.	*Gonia chinensis* (Wiedemann, 1824)	Goniini	Exoristinae	Tachinidae	*Anomis flava*	March-August
21.	*Lophosia bicincta* (Robineau-Devoidy,1830)	Cylindromyiini	Phasiinae	Tachinidae	Unknown	June-August
22.	*Eurithia indica* (Lahiri, 2003)	Ernestiini	Tachininae	Tachinidae	Unknown	June-August
23.	*Therobia abdominalis* (Wiedmann, 1830)	Ormiini	Tachininae	Tachinidae	Unknown	June-August
24.	*Actia mimetic* (Malloch, 1930)	Siphonini	Tachininae	Tachinidae	*Eutectona machearalis*	April-October
25.	*Tachina fulva* (Walker, 1853)	Tachinini	Tachininae	Tachinidae	Unknown	April-October
26.	*Tachina subcinarea* (Walker, 1853)	Tachinini	Tachininae	Tachinidae	Unknown	April-October

known in india. It seems that Indian fauna of Tachinidae is very rich and further studies are needed. The Tachinids were mostly associated with lepidopterous pests and noticed their supportive in control of lepidopterous pests.

Acknowledgement

Authors are thankful to UGC, New Delhi for financial assistance to Major Research Project F-37-334/2009 (SR)/37-1/2009 (MS) (SR) and ShivajiUniversity for providing facilities.

References

Crosskey, R.W., 1976. A taxonomic conspectus of Tachinidae (Diptera) of the oriental region. *Bull. Brit Museum (Nat. Hist.) Ent. Suppl.*, 26: 357.

O'Hara, James, 2008.World genera of the Tachinidae (Diptera) and their regional occurrence version, 40.

Lahiri, A.R., 2003. Diptera: Tachinidae in fauna of Sikkim, State fauna series (part–3) 387–399. *Zool. Suvey India*, Kolkata.

Lahiri, A. R., 2006.Diptera: Tachinidae in fauna of Nagaland, State fauna series 12, 199–211. *Zool. Suvey India*, Kolkata.

Narayanaswamy, K.C., Devaiah, M.C and Govindan, R., 1933. Laboratory studies on ovipositional preference and biology of Uzifly, *Exorista bombycis* on some species of Lepidoptera. *Recent. Adv. Uzifly Res.*, pp. 43–48.

Sathe, T.V., 2012. Diversity of Tachinids (Diptera; Tachiniae) from Western Maharashtra. *IJJP*, 5(2): 368–370.

Sathe, T.V. and Jadhav, A.D., 2001. *Sericulture and Pest Management.* Daya Publishing House, pp. 1–147.

Chapter 19

TACHINIDS IN PEST MANAGEMENT:

(B) Biology of *Eucelatoria bryani* Sabrosky (Diptera : Tachinidae): A Larval Parasitoid of *Helicoverpa armigera* (Hubn.) (Lepidoptera: Noctuidae)

Nilam Shendge and T.V. Sathe

Department of Zoology, Shivaji University, Kolhapur – 416 004, Maharashtra, India

ABSTRACT

Helicoverpa armigera (Hubn.) (Lepidoptera: Noctuidae) is polyphagous insect pests. It cause damage to more than 180 host plants and difficult to control with conventional pesticides. Therefore, as a natural enemy of *H. armigera* biology of *Eucelatoria bryani* Sabrosky (Diptera: Tachinidae) has been studied under laboratory conditions (25±1°C, 70-75 per cent R.H; 12hr photoperiod). The life cycle from egg to adult was completed within 15 to 18 days on *H. armigera*. Incubation, larval and pupal periods were 3-4days, 7-8days, and 5-6days respectively. The present work will be helpful for mass rearing of the parasitoid and further, in biological control of *H. armigera*.

Keywords: *Eucelatoria bryani, Biology, Parasitoid, Helicoverpa armigera, Pest.*

Introduction

According to Doutt, (1959) thousands of parasitic wasps occur throughout the World, possessing complex and fascinating biologies and frequently determine population densities of pest insects. Hence, they have tremendous economic importance in pest control strategies (Coppel and Mertins, 1977). The parasitoids are widely scattered in orders Hymenoptera, Diptera, Lepidoptera, Strepsiptera and Hemiptera. However, order Diptera plays very important role since they parasitizes Lepidopteran, Orthropteran, Coleopteran and other pests. From Diptera the family Tachinidae is exclusively parasitic in nature and useful in biological control of insect pests as biocontrol agents. *Helicoverpa armigera* (Hubn.) (Lepidoptera: Noctuidae) is serious polyphagous pest of several agricultural, horticultural and forest crops. Hence, hoping its utility in biological control programme of *H. armigera*, biology of this parasitoid has been studied under laboratory conditions.

Review of literature indicate that biology of tachinds have been attempted by Datta and Mukharjee. (1978), Devaian *et al.* (1993), Thomson (1944), Shekharappa *et al.* (1988), Sankaran and Nagaraja (1979).

Materials and Methods

For initial culture of Tachinid fly *H. armigera* from agricultural fields of Kolhapur were collected and rearing on natural food plants in the laboratory for screening parasitoids. Screened parasitoids were kept individually in test tubes and allowed to mate in test tubes in pairs (1 male and 1 female) and mated females were exposed to 3rd instar larvae of *H. armigera* in glass cage 25×23×25cm (in length, width and height) for 24hr for parasitization and then parasitized *H.armigera* larvae were separated and reared individually in small containers for avoiding cannibalism by providing its natural food, either gram leaves/turpods. Observations were noted on life cycle of the parasitoid. Sufficient number of larvae were parasitized, were collected and

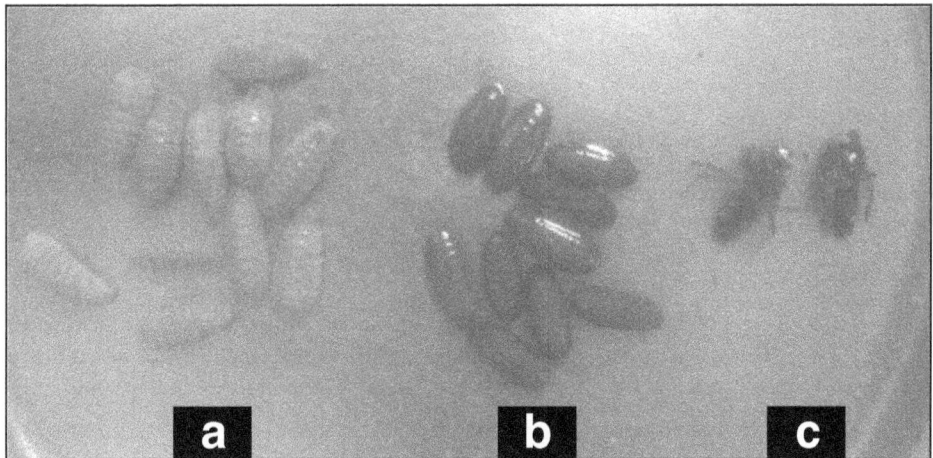

Figure 19.1: Tachinidfly–
(a) Maggots; (b) Pupae; (c) Adults

dissected every day in ten groups for the development of immature stages and instars and further immature stages have been identified along with the duration taken by each stages and life cycle duration was finalized.

Results

The parasitoid showed 4 district stages of its life cycle. *viz.* Egg, Larva, Pupa and Adult.

Egg

Newly deposited eggs were whitish, thin walled, elongated and tapering towards both ends (hymenopteriform) (Figure 19.1). Incubation period was 3-4 days.

Instars

There were 3rd instars in the parasitoid life cycle.

1st Instars

First instar was opaque, translucent and with quadrate head,3 thoracic and seven abdominal segments. The instar was characterized by small tail. No leggs and spiracles were observed in this instar. This stage lasted for 2-3 days. The larva consumed internal tissues of the host larva.

2nd Instars

Second instar was cylindrical, whitish, more or less straight flatish and with 13 well defined segments. Spiracles were observed on the abdominal segments of the larva. This stage was characterized by small narrow and lasted for 3-4 days. The larva consumed internal tissue of the host larva. The larva was broader at posterior.

3nd Instars (Figure 19.1a)

3nd instar larva was with typically shortened and anterior tapered towards both end and was with well defined 13 body segments. Mandibles well developed. Tail was absent in this form. Six pairs of spiracles were visible on abdominal segments. The opaque white coloured larva come out of the host body for spinning cocoon.

Cocoon (Figure 19.1b)

This stage lasted 3-4 days. The parasitoid cocoon was brownish and barrel shaphed and with well segmented. Third instar larva spin the cocoon by coming out of the host body. Within half an hour. Pupal stage lasted for 5-6 days.

Adult (Figure 19.1C)

Adults were blackish with 4 grayish longitudinal strips on thorax and with chaetotaxy on thorax and abdomen. The fly measured for 3mm to 3.5mm in body length and was characterized by a pair of halters. The ovipositor was very short.

The life cycle from egg to adult was completed within 15-18 days at laboratory condition (25 ±1C, 70-75 per cent R.H. and 12hr photoperiod).

Discussion

Tachinids are of two categories. In one case they lay eggs directly on the host body and such eggs are larger in size. Another type of tachinids lay their eggs on host plant and their eggs are thus smaller in size. In this category, the eggs entered into host body through mouth or alimentary canal and after hatching the eggs, the larvae started eating internal tissues of the host body.First category tachinids lay their eggs into host body. The parasitoid larva moults three time in host body. The third instar, which is fully matured break the body wall of the host and come out and spin the cocoon. While breaking body wall of host and coming outside the host body, the host dies on the spot. Thus, helps controlling pest species.

Devaian *et al.* (1993) studied the life cycle of *Exorista bombycis,* and reported that the fly has completed 5-8 generations in a year, and one generation was completed within 15.4 to 50 days depending on climatic conditions. Recently, Sathe and Jadhav (2001). studied the same parasitoid under laboratory conditions (25±1°C,753 per cent R.H. and 18:6hr photoperiod) are reported 32.9 days as life cycle period from egg to adult formation. Mani (1985) reported 38 days for completion of the cycle was faster than *E. bombycis.* Which is useful attribute for mass rearing of this species. The present work will be helpful for mass rearing of *E. bryani.*

Acknowledgement

Authors are thankful to UGC, New Delhi for financial assistance to Major Research Project F-37-334/2009 (SR)/37-1/2009 (MS) (SR) and Shivaji University for providing facilities.

References

Beeeson, C.F.C. and Chatterjee, S.N., 1935. The biology of Tachinidae (Diptera). *Forest Records,* 1L 84.

Datta, R.K. and Mukherjee, P.K., 1978. Life history of *Tricholyga bombycis* (Diptera: Tachinidae), a parasite of *Bombyx mori* L. (Lepidoptera: Bombycidae). *Ann. Ent. Soc. Am.,* 71: 767–770.

Devaiah, M.C., Govindan, R. and Narayanswamy, K.C., 1993. Life table of uzifly*Exorista bombycis.* In: *Recent Advances in Uzifly Research.* Proc. Nat. Semi. Uzifly and its Control, 1992, p. 1 –12.

Mani, M., 1985. Age specific fecundity and rate of increase of *Eucelatoria bryani* Sabrosky on *Helicoverpa armigera* (Hubn.). *Indian J. Ent,* 47(2): 163–168.

Sathe, T.V. and Jadhav, A.D., 2001. *Sericulture and Pest Management.* Daya Publishing House, Delhi, pp. 1–167.

Shekharappa, K.J. and Lingappa, S., 1998. Studies on the Supperparasitism in *Eucelatoria bryani* Sabrosky, a larval parasitoid of *Heliothis armigera* (Hubn.). *Entomon,* 13(3 and 4): 201–206.

Chapter 20

WEEVILS IN BIOLOGICAL CONTROL OF WEEDS

T.V. Sathe* and P.V. Khairmode

*Department of Zoology, Shivaji University,
Kolhapur – 416 004, Maharashtra, India*

Introduction

Weevils belong to the family Curculionidae of Order Coleoptera. Most of the weevils are phytophagus. However, many species found feeding on various weeds and act as biological agent of them.Therefore, various cases of biocontrol programs of weeds at global scenario are represented in this chapter.

1. Weevils in Weed Control

Smith (2010) used an aquatic weevil, *Euhrychiopsis lecontei* (Dientz) for biological control of *Eurasion watermilfoil* (*Myriphyllum spicatum*) in Michigan's Les Cheneaux Islands. The density and vigor of *E. watermilfoil* (EWM) was quantified for two years following the introduction of *E. lecontei* in 2007. This has been resulted in reduction of EWM density and re-growth of native macrophytes. Fourteen months following *E.lecontei* introduction, a decrease of eighty-five percent in EWM to native *E.lecontei* growth was observed at both test sites. Native macrophytes began to grow in native. *E.lecontei* population was a probable factor in successful EWM control in this project. EWM present has been significantly reduced. However, native macrophytes are again occurring and recreational boaters are now transit engaged in portion of the bay that

* E-mail: profdrtvsathe@rediffmail.com

were minimally navigable. Probably, reintroduction of weevils and resurvey of weed is essential part of the end success of the project.

Method used by Sloan (2009): Thirteen thousand weevils were planted at site-1 and two thousand weevils were planted at site-2. A third, distant site in the same bay was used as an untreated control (UTC) site (Sloan, 2009). Site-1 was located in 3-4 ft of about ten meters east of the Boat School Channel. Site-2 was in 2-4 ft of water forty meter east of Dock cribs at the old Shoberg resort.

A survey was conducted at each site to identify native macrophyte species present prior to planting weevils estimated by sampling a specific number of specimen based on a statistical model.

After fourteen months post weevils have been introduced which resulted only five percent of the initial EWM density reduction at site-1. After twenty- six months the EWM density remained low at about 10 percent of that initial observed. Site-2 pattern was similar in that less than fifteen percent of the initial EWM crop was present at fourteen months and after twenty – six months less than five percent of the EWM density was recorded at beginning of the study.

2. Classical Biological Control of Malaleuca Weeds by Malaleuca Weevil

According to Cuda *et.al* (2010), Malaleuca, *Melaleuca quinquenervia* (Cav.) (Myrtaceae), is an invasive woody plant in Australia, New Guinea, and Soloman Islands. Malaleuca, also known as the paper bark tree. In 2008, Melaleuca infested over 110,000 hector of wetland ecosystem on Florida. Therefore, biological control practice has been used to manage Melaleuca in South Florida. Biocontrol methods have been successfully integrated to provide the most effective control of this weed.

As a Classical biological control, introduction of host specific natural enemies was investigated as a possible long-term solution to the Melaleuca problem. Five insects have been released in addition to the Melaleuca snout weevil *Oxyopus vitiosa* (Pascoe) (Coleoptera: Curculionidae).

Melaleuca weevils are grey in colour and 6 to 9 mm in length and somewhat cryptic in appearance, usually indicated by the characteristic feeding damage that consist of holes or gouges chewed into the buds, leaves and stems.

3. Biological Control of Paterson's Curse with Crown and Root Boring Weevils

Paterson's curse, *Echium plantagium,* is a noxious weed of European origin that now occurs in most states of Australia and is mainly a problem in pastures, on roadsides and in degraded and disturbed areas. It has reduced the agriculture productivity at a greater extent by competing with one nutritional pasture plant and because it is toxic to livestock when ingested continuously.

The crown boring weevil is 3.5 to 4 mm long. It is brown and white with a characteristic pattern. The root boring weevils are slightly larger, 4 to 5 mm, and have a light brown body covered with fine, pale lines. Both beetles have very hard bodies, a long snout (the rostrum), with mouth parts located at the tip, and elbowed, club – shaped antennae that can be folded against the snout.

Damage by Weevil

Larvae damage the crown and root making plants less competitive and reducing their ability to produce seeds. Adult chew holes in leaf blades causing "shot hole" damage and also feed on leaf stalks. The plant may from calluses where the stalks have been attached.

Release of Weevils

The crown weevils have been released at over 1000 sites in Australia since 1994 and have been redistributed from 1996 onward. The root weevil was released in early 1996. It has proved to be much harder to rear, so few releases have been made. Thus, this weevil proved its efficiency as biocontrol agent.

4. Echhornia Weevils *Neocherina bruchi*

At Bangalore, India liliacy weed has been controlled by using Echhornia weevils *Neocherina bruchi*. Attempts have been also made to control Kendal (Local Name) *Echhornia crassipes* liliacae aquatic weed in Rankala Talav, Kolhapur city with the help of Echhornia weevil, *Neocherina bruchi* but no success has been achieved for the control of this weed in Kolhapur, India.

5. Mexican Beetle (*Zygogramma bicalrota*) for Control of Parthenium

In India congress grass, Parthenium has been successfully controlled by using Mexican beetle (*Zygogramma bicalrota*) (Coleoptera: Chysomelidae). A Mexican beetle emerged at the beginning of Monsoon season. Newly emerged pair (male and female) mate for 3 to 4 days and mated female start egg laying on the leaf surface or under the surface of leaves either singly or in clusters of 4-5 eggs. The eggs hatch within 4 to5 days. Single mated female can lay about 65 eggs per day with a total of 800 to 900 in a whole life span. The larva moults for 3 times for pupation. The larval period is 10 to 15 days and pupal period is 8 to 10 days. The beetle goes in diapause from November to May/June.

These beetles are mass reared in the laboratory on the leaves of Congress grass and released with 500 individuals (adults) per hectare for control of above grass. The beetles may be collected from the natural habitat where they are abundant and released in target area for control of Parthenium grass. Parthenium or congress grass is very troublesome weed in India which is scattered in almost every state in India and affecting status of several agricultural crops. Hence, its control is the need of the day.

Acknowledgement

Authors are thankful to Prof. and Head, Dept. of Zoology, Shivaji University, Kolhapur for providing necessary facilities for this work.

References

Balciunas, J.K., Burrows, D.W. and Furcell, M.F., 1994. Field and laboratory host ranges of the Australian weevil, *Oxyops vitiosa* (Coleoptera: Curculionidae) a potential control agent for the paperbark tree, *Melaleuca quinquenervia. Biol. Control*, 4: 351–360.

Center, T.D., Drav, F.A. and Vandiver Jr., V.V., 2000. Biological control with insect : the Melaleuca snout beetle. EDIS website http: //edis. Ifas. Edu/AGO22. Florida Cooperative Extension Service, IFAS, University of Gainesvill, FL.

Sathe T.V., 2004. *Vermiculture and Organic Farming*. Daya Publishing House, New Delhi, pp. 1–120.

Sloan, J., 2009. Les Cheneaux Island Watershed Council Middfoil Program Summery. Enviroscience, Inc.

Smith, R.A., 2008. Status report on the impact of planting weevils, *Eubrychiopsis lecontei* (Deitz) in Cedarville Bay to combat Eurasian watermilfoil (*Myriphullum spicatum*) during the summer of 2007.

Chapter 21

PISCES FOR MOSQUITO CONTROL

T.V. Sathe and R.V. Kawane

Department of Zoology, Shivaji University,
Kolhapur – 416 004, Maharashtra, India

Introduction

Several fishes have little food value but have immense utility to public health because of their larvivorous feeding habits. These fishes prefer insect larvae such as mosquitoes, red worms, gnates, etc over other food. The larvicidal function of these fishes has brought them to prominence in epidemiology. Man has learnt to use them as one of his several measures to control a number of serious diseases such as malaria, filaria, dengue, chikungunya, yellow fever etc. spread by mosquito species. Man's war against these dreaded diseases, is a war against their vector insects for eradication of diseases. Control of vector insects is becoming difficult day by day because of development of resistance to pesticides. Hence, biological control is the only alternative left to keep a check on the vector insects. Destruction of larvae eventually causes a rapid reduction and control of the adult insect population. Since long larvicidal fishes have been utilised in mosquito control. Tropical countries have swamps, marshes, etc. for breeding places for the vector insects and therefore becomes a potential source of the menace of the diseases. Mosquitoes play a very important role in the spread of above diseases.

Larvicidal fishes directly feed upon mosquito larvae, and thus their introduction in water bodies where mosquitoes breed results in a very effective control of the mosquitoes. For selection of the larvicidal fishes for a control of mosquitoes preference is given to such fishes which shows following features:

1. Small in size
2. Breed freely
3. Have little food value
4. Hardy enough to withstand the difficult ecology of swampy waters
5. Its movement should be swift enough to escape from its enemies
6. Fish should be carnivorous and its main diet should be insect larvae
7. Fish should be surface feeder.

Mosquito borne diseases is chronic problem in almost all tropical and subtropical countries of the world. Mosquitoes transmit pathogens causing some of the most life - threatening and debilitating diseases of man, like malaria, yellow fever, dengue fever, chikungunya, filariasis, encephalitis, equine diseases, etc.

Many pesticides have banned by Environmental protection agencies or placed severe restrictions on their use which were formerly used in mosquito control programmes. Therefore, now only fewer adulticidcs are available for use against mosquitoes. Further, manufacturing of specific insecticides have withdrawn by pesticide companies due to the high cost of carrying out the additional tests as per the government norms. The fact that the production of crop pesticides for the agricultural market is much more lucrative than mosquitoes controlling pesticides. Secondly, the harmful effects of chemicals on biological agents as well as on non-target populations, and the development of resistance to these chemicals in mosquitoes along with recent resurgence of different mosquito borne diseases have stimulated to find out an alternative for pesticide use and for developing simple sustainable method of mosquito control. The eradication of mosquito using adulticides is not a prudent strategy since they are with human habitation and can easily escape from control measures.

Introduction of larvivorous fishes for biological control of mosquitoes was important malaria control programmes in the 20th century, particularly in urban and periurban areas of developed and developing countries. In the integrated control methodologies in which both pesticides and fishes or other biotic agents have their own roles. Biological control refers to the introduction or manipulation of organisms to suppress vector populations. The mosquito populations are suppressed naturally through predation, parasitism and competition by BCAS. Biocontrol agents like larvivorous fishes (*i.e.*, those that feed on immature stages of mosquitoes) are being used extensively all over the world since the early 1900s (pre DDT era).

Biocontrol agents like dragonflies, fishes, planarians, hydras, fungi, bacteria, protozoans and nematodes are used for mosquito suppression throughout the world. However, fishes have proved tremendous potential in mosquito control programme. During the pre DDT era, mosquito vector management was mainly by environmental management, pyrethrum space spraying, and use of Paris green, oiling with petrol products and introduction of larvivorous fishes. High potential larvivorous fish *Gambusia affinis* was purposely introduced from its native Texas (Southern USA) to the HawaiinIslands in 1905. Later in 1921, it was introduced in Spain and then in Italy during 1920s. *G. affinis* was also introduced in 60 other countries for control of mosquitoes. A larvivorous fish, *Poecilia reticulata*, a native of South America, was introduced for malaria control in UK, India and many other countries in 1908. Review of literature indicates that DDT was introduced for use as indoor residual spraying for malaria control around the mid - 1940s which has resulted in gradual decline in the use of concepts of environmental management and biological control methods. However, Russia continued with a few programmes without DDT. Attention was directed to eradicate mosquitoes using synthetic insecticides until insecticide resistance began in mosquitoes around 1950 but, WHO changed its strategy of malaria eradication by spraying houses with synthetic insecticides in favour of the more realistic one for the control of mosquito populations in the larval stages in 1969 as past DDT era. Biological control of mosquitoes has tremendous importance for keeping environment healthy. A biocontrol agent to be used in control programme should have following features:

1. Potential for unintended impacts
2. Self-replicating capacity
3. Climatic compatibility
4. Capability to maintain very close interactions with target prey populations.
5. Predator should possess extraordinary search efficiency irrespective of the illuminated situation in response to the emergence of prey.
6. Predator's adaptability to the introduced environment and overall positive interaction with indigenous organisms. It is also expected that implementer should have a sound knowledge of predator's prey selective patterns, and particularly mosquito larvae selection in the presence of alternate natural prey.

Criteria for Species Selection

Larvivorous fishes feed on immature stages of mosquitoes. According to Job (1940) larvivorous fish must be small, hardy and capable of getting about easily in shallow waters among thick weeds where mosquitoes find suitable breeding places. Fish must he drought resistant and capable of flourishing in both deep and shallow waters as well as living in drinking water tanks and pools without contaminating the water. The fish should have the ability to withstand rough handling and transportation for long distances. The fish must be prolific breeders and must have shorter span oflife cycle. The fish must breed freely and successfully in confined waters and should be surface feeders and carnivorous in habit and should have a predilection for mosquito larvae even in the presence of other food materials. The fish should not he brightly coloured or attractive. The fish should be compatible with the existing fish life in that environment. The fish should have no food value.

The parameters given above can not seen in a single species. Therefore care must be taken that as much as many of the above qualities be fulfilled by the larvivorous fish to be used in mosquito control programme.

On the basis of mosquitocidal activity, the larvicidal fishes are classified into following categories:

1. Surface Feeders

The species which are surface feeders. *e.g. Aplochielus* and *Gambusia* fall in this catagory. These fishes areideal larvicidal fishes, hence are highly efficient for malaria control programme. *Oryzia, Aphaniui*, and *Horaichthys* are also surface feeders but are relatively less efficient.

2. Sub-Surface Feeders

These species are sub surface feeders. *e.g. Dano, Rasbora, Esomus* and *Carassius.* These are easy to raise but are larvicidal to a considerable extent.

3. Column Feeders

Column feeder fishes refers to *Punctius, Amhassis, Anabus, Colisa, Macropodus, Badis* and *Therapon* when they get chance they can feed on larvae of mosquitoes.

4. Large Sized Food Fishes

Large sized fishes like *Catla, Lebeo, Mugil,* etc. are also quite effective for control of mosquito larvae.They feed on mosquito larvae and suppress the population of mosquitoes.

5. Predator Fishes

Predatory fishes like *Wallago, Channa, Notopterus* and *Mystus* feed upon mosquito larvae and acts as biocontrol agents for controlling mosquito population. Adults are predacious upon other pond fishes.

For the actual mosquito population control the systematic planning is essential. The larvicidal fishes can be introduced into various water bodies where breeding of mosquitoes occur. However, the dense vegetation, prior to introduction of fishes, should be cleared. Fish predators be completely removed from introduction areas. The recommended dose for mosquito control is two females and one male for *Aplucheiltis* and one male and one female for *Gambusia* in a area of 100 sqm will give good control of mosquitoes.

Fishes Used in Mosquito Control

In general following types of fishes are used in biological control of mosquitoes.

1. Efficient surface feeders : *Aplochelies, Gambusia* etc.
2. Non efficient surface feeders : *Poecilia (Lebistes), Oryzia, Aphanius,* etc.
3. Sub surface feeders : *Danio, Rasbora, Amblypharyn godonmola,* etc.
4. Colum feeders : *Anabus, Channa, Colsia, Punctius,* etc.
5. Fry of carps and mullets:
6. Predatory fishes : *Notopterus, Channa, Mystus, Wallago,* etc.

There are several mosquitovorous species of fishes which are indigenous as well as exotic. Exotic and indigenous species which feed on mosquito larvae are tabulated in Table 21.1.

Table 21.1: Some Important Larvicidal Fishes Used in India.

Sl.No.	Common Name	Scientific Name	Status
1.	Cichlid fish	*Gambusia affinis*	Exotic
2.	Cichlid fish	*Lebistes reticulata*	Exotic
3.	Cyprinid	*Carassius carassius*	Exotic
4.	Minnow	*Aplocheilus lineatum*	Indigenous
5.	Minnow	*Oryzia melanostigma*	Indigenous
6.	Minnow	*Aphaniits dispar*	Indigenous
7.	Minnow	*Horaichthys*	Indigenous
8.	Minnow	*Amblypharyngodon*	Indigenous
9.	Cyprinid	*Rasbora daniconius*	Indigenous
10.	Cyprinid	*Esomus danaricus*	Indigenous
11.	Cyprinid	*Punctius sophore*	Indigenous

In India, Malaria is one of the major scourges. Its prevention and cure have attracted the attention of the Medical and Public Health authorities for well over quarter of a century. There are several concepts for the problem of the prevention of Malaria but, from the time of the discovery by Sir Ronald Ross that the Anopheline mosquito was the carrier of the malarial parasites, it has been the aim of the authorities to control the incidence of Malaria by controlling the breeding of the carrier-mosquitoes in various ways. It has been reported that many species of Indian freshwater fishes have a special preference for mosquito larvae as food, and that their introduction into tanks, ponds and wells is the way to control the mosquito population in the neighbourhood of human dwellings. Therefore, medical men gave their attention to the fishes of the aquatic areas in which mosquitoes breed but, found themselves in difficulties in the identification of the fishes. The only standard literature on the Fishes of India are none too easy to refer to, are burdened with a mass of technical details. Therefore resulted in often wrong identification of species or sometimes not at all. For the medical man, who is a field-worker interested in the control of Malaria by using larvicidal fishes, needs a simple guide for identification of freshwater fishes in India. On this background, the present work will add great relevance.

Many carnivorous fishes feed upon the zooplanktons (cladocerans and rotifers) while, herbivorous fishes feed upon phytoplankton's (algae, aquatic weeds etc.). According to Atkin many fishes consume the insect larvae as favourite food. Mosquito borne diseases like filaria, malaria, yellow fever, dengue fever, chikungunya etc. can be eradicated with the help of larvicidal fishes by gradually decreasing mosquito population in natural environment. Control of mosquitoes through their natural enemies is called as biological control. Biological control now regarded as one of the most potent tool for destroying mosquitoes without causing any harmful effects on human health and food chain.

Indigenous Fishes

1. Dispar Topminnow

Order–Cyprinodontiformes

Family–Cyprinodontidae

Genus–*Aphanius*

Species–*dispar* (Ruppell) 1828

Distribution–Ethiopia, India, Pakistan, Red sea, etc.

Marks of Identification

1. It measures about 7 to 8 cm in body length.
2. Body slender, spindle shaped.
3. Single dorsal fin originated from the region closer to tail.
4. Lateral blackish, grayish stripes present dorsally. It is a blackish, ventrally it is a whitish.
5. Caudal fin with blackish patch to the basal region, homocercal tail fin.

Special Features

1. Live in fresh and brakish water and breed freely.
2. Feed on mosquito larvae specially *Culex quinquefasciatus* Say, *Anopheles arabiensis, Anopheles gambie, Anopheles culicifacies adanensis,* etc.
3. It can breed in drains and polluted water bodies, stagnant water bodies, wells and cesspools, etc.

Utility

It is used for biological control of above said mosquitoes. In natural habitat 3 fishes per squre meter of water are released for successful control of mosquito larvae. This is possible in natural habitat particularly in shallow channels. In man made habitat like containers it can act successfully against *C.quinquefasciatus.* It can also control above mosquito larvae successfully in stagnant water containing organic pollution. *A. dispar* can also effectively suppress the mosquito larvae of the species *An. arabiensis* and *An. gambiae*in wells, cisterns, barrels and containers. In the urban area of Euthopea, the population of *A. culicifacies adanensis* have been effectively suppressed by the fish *A.dispar.* It has been also noticed that in cisterns *A. dispar* effectively controls *An. arabiensis* and *An. gambiae* mosquitoes in the urban area of Djabouti. In India, it should be used on large scale against the mosquito control as a part of malaria and filaria control measures.

2. Dwarf Panchax

Order–Cyprinodontiformes

Family–Aplocheilidae

Genus–*Aplocheilus*

Species–*blockii* (Arnold)1911

Distribution–India. In India it has been reported from eastern coast from Madras north wards.

Marks of Identification

1. It is elongated with transparent dorsal fin close to tail region.
2. It measures about 9 cm.
3. Anal fin is close to tail region.
4. Scales are silvery. It is with silvery and whitish, bluish lateral bands.
5. Eyes are black, small sized ringed with silvery shining markings.

Special Features

It is fresh water breeder hence found in stationary and shelter water of tanks, small streams and rivulets overgrown with thick vegetation. It can also breed in wells and abounded water bodies.

Utility

It can be introduced for control of mosquitoes in overhead tanks. Ornamental pools, stream belts and margins, wells and reservoir. It can effectively control *Anopheles stephensi* larvae.In the costal belt of Goa the fish suppressed 75 per cent population of mosquitoes. It can control *Aedes albopictus* (Skuse, 1894) in tanks and bigger cisterns and barrels in India and prevent the spread of chikungunya in the region.

3. Malabar Killie

Order–Cyprinodontiformes

Family–Aplocheilidae

Genus–*Aplocheilus*

Species–*lineatus* (Valencinnes)1846

Distribution–India and Sri Lanka. In India, it is widely distributed in peninsular region.

Marks of Identification

1. The fish measures about 10 cm in body length.
2. It is elongated with transparent dorsal fin close to tail region.
3. Anal fin is close to tail region.
4. Scales darkish silvery with silvery and whitish, bluish lateral horizontal bands.
5. Eyes are black, small sized ringed with silvery shining markings

Special Features

It is commonly found in tanks, paddy fields, canals, and tidal water, pools, streambeds, margins and marshes in rural areas etc.

Utility

In man made habitats it is very potential biocontrol agents of *Aedes aegypti*. It can breed in storage tanks, cisterns and barrels therefore it is quite useful.

4. Panchax Minnow

Order–Cyprinodontiformes

Family–Aplocheilidae

Genus–*Aplocheilus*

Species–*panchex* (Hamilton-Buchanan) 1822

Distribution–India (Bengal, Bihar, Orissa, Assam, Punjab, Uttar Pradesh, Madhya Pradesh, Rajasthan), Sri Lanka(Malaya)Myanmar, Thailand, and Indonesia.

Marks of Identification

1. The fish measures about 9 cm in body length.
2. It is elongated with transparent dorsal fin close to tail region.
3. Anal fin close to tail region.
4. Scales silvery with silvery and whitish, bluish lateral bands.
5. Eyes are black, small sized ringed with silvery shining markings

Special Features

It is commonly found in wells, marshes, lagoons and polluted storm water drains and any other stagnant water bodies. The species is quite hardy and active and inhabits clear shallow fresh and brackish water at low altitudes.

Utility

It is effective biocontrol agent of *Culex* mosquitoes which transmit Filaria. It is also effective against *Aedes* mosquito larvae, it is very effective biological agents of *An. culicifacies* in breeding habitats such as sluggish streams, rain water pools, irrigation channels, river bed pools, borrow pits, cemented tanks, swimming pools, and paddy fields. In brakish water it can control *An. sundaicus*. It is also effective biocontrol control agent for *Culex quinquefasciatus* in wells, manure pits, disused wells, septic tanks, cess pools, drains, etc.

5. Dwarf Gourami

Order–Perciformes

Family–Osphronemidae

Genus–*Colisa*

Species–*lalia* (Hamilton-Buchanan) 1822

Distribution–Northern India, Assam, Bengal, Bihar and Uttar Pradesh.

Marks of Identification

1. The fish measures about 5 cm in body length.

2. It is broadly oval in shape and with single dorsal fin covering basal part of tail fin.
3. Anal fin close to tail region.
4. Tail fin is somewhat rounded, Mouth is superior.
5. Eyes are black, small sized, ringed with silvery shining markings

Special Features

It inhabits slow moving streams, rivulets and lakes with dense vegetation. Suitable for water bodies where carnivorous food fish are present. It can breed in lakes, tanks, *etc.*

Utility

It is useful for introduction in clear water, slow moving stream with grassy margins and shallow earth wells on seepages for control of Anopheles mosquito larvae which acts as vector for malaria. In natural habitat it can effectively control *An. (c) annularis.* It can be released in stagnant water, tanks, dead rivers, borrow pits and rice fields for effective control of above said mosquito species.

6. Giant Gourami

Order–Perciformes

Family–Osphronemidae

Genus–*Colisa*

Species–*fasciatus* (Schn.)

Distribution–India, Pak, Myanmar, etc.

Marks of Identification

1. It measures about 5 inches in length.

Utility

It control the mosquitoes *An.(c) annularis* and *Mansoniodes* indiana Edward causing malaria and filaria respectively. More than 100 fishes can control mosquitoes effectively in wells.

7. Pholi Fish

Order–Perciformes

Family–Notopteridae

Genus–*Notopterus*

Species–*notopterus*

Distribution–India : Tamil Nadu, Maharashtra etc.

Marks of Identification

1. Its body is silvery white and shows fine grey spots on the back.

2. It measures about 2 -45cm in body length but can attain about 60 cm.

3. It is developed in large laties and reservoirs.

Utility

It control the mosquitoes *Culex* spp., *Anopheles* spp., *Aedes* spp., it is used as biocontrol agents for mosquito larvae.

8. Spiketailed Paradise Fish

Order–Perciformes

Family–Osphronemidae

Genus–*Mucropodus*

Species–*cupanus* (Valenciennes)1831

Distribution–Eastern India. Sri Lanka, Western, Myanmar, Malay Peninsula and Sumatra.

Marks of Identification

1. The fish measures about 7.5 cm in body length.

2. Mouth is superior.

3. Single dorsal fin, ventral fin is elongated, pelvic fins are paired it is near ventrally.

4. Tail fin is ovate shaped.

5. Body is fusiform.

Special Features

*M. cupanus*is good larvivorous fish thriving both in fresh and brackish waters of the low lands, also found in ditches, paddy fields and shallow waters.It is also suitable for brackish waters, marshes, lagoons, polluted canals and ditches. It breeds freely in stagnant waters and is tolerant to low content or even deficiency of oxygen.

Utility

It is effective biocontrol agent of *Culex, Anopheles* and *Aedes* mosquito larvae. A significant results have been reported against *Cx. fatigans* by Mathwan *et al.* (1980).

9. *Oryzia melanostigma*

Distribution–India

Marks of Identification

1. It is small size fish.

2. It is found in fresh water and shallow water.

3. It is delicate and can not be transposes easily.

Utility

It is very useful biocontrol agent of mosquito larvae.

10. *Danio rerio*

Distribution–India

Marks of Identification

1. It measures about 5 cm in length.
2. It is with upturned mouth.

Utility

Acts as biocontrol agents for mosquitoes.

11. *Mugil*

Order–Perciformes

Family–Mugilidae

Genus–*Mugil*

Species–*macrolepis* (Smith)

Distribution–India: Bombay coast, Gulf of Mannar, along peninsula, TamilNadu, Maharashtra.

Marks of Identification

1. It measures about 80-90 cm in body length and are with grey body and head contain cycloid scales.

Utility

Young ones of this species are effective biocontrol agents of mosquito larvae. It is commonly known as mullets. Some species live in essuarine and fresh waters. The young ones feed on mosquito larvae.

12. *Esomus danaricus*

It measures about 13 cm in body length. It occurs in the fresh water ponds, ditches and streams of India. It can breed effectively in above habitats.

Utility

Effective for *Anopheles* and *Aedes* mosquitoes.

13. *Barilius vagra*

This small fish is found in hill streams.

Utility

It can effectively control the mosquito larvae by feeding upon them.

14. *Punctius* spp.

Several species of the genus *Punctius* are found in fresh water.They are commonly called carp minnows. These hard fishes are small sized and have no food value.

Utility

They are used for control of mosquito larvae in India.

15. *Resbora daniconius*

It measures about 3-13 cm in body length. It is free breeder in fresh water of India.

Utility

It controls mosquito larvae very effectively by feeding upon them.

16. *Anabus testudineus*

This large sized fish measures about 30 cm in body length and is a very hardy fish and has economic importance. It breeds in oxygen depleted waters of ponds, sewage and canals.

Utility

It is useful for mosquito control. They feed on larvae of *Culex,* and *Anopheles* and acts as very good biocontrol agents.

17. *Etroplus*

It is found in brackish water and there are two species in India. Adults feed upon algae.

Utility

The young ones of this species are potential mosquito larvae feeders and used in biocontrol of mosquitoes.

18. *Badis*

It is measures about 5-8 cm in body length and is beautifully coloured fish. Breed in fresh water.

Utility

This fish is effective biocontrol agents of mosquito larvae of *Aedes* and *Anopheles* species.

19. *Oxygaster*

Several species are reported under the genus *Oxygaster* from India. These fishes are commonly called as chilwa.

Utility

The smaller varieties feed on mosquito larvae. These delicate, large fishes are used as food, used in biocontrol of mosquito. In addition of above species *Laubuca, Ambasis, Therapon jarbus* etc. are also found effective in mosquito control in India.

Exotic Species

1. Top Minnow

Order–Cyprinodontiformes

Family–Poeciliidae

Genus–*Gambusia*

Species–*affinis* (Bairdand Girard) 1853

Distribution–India, USA (native), Thailand, Italy, etc.

Marks of Identification

1. It measures from 2.5–3"cm in length in males and 6 cm in females.
2. Body is elongated.
3. Mouth is upturned and head is small.
4. Different colour variation.

Special Features

Top minnow (*Gambuxia affinis*): usedfor mosquito control for several species. *Gambuzia* (*G. affinusholbrooki*) can be kept in the aquarium and used against mosquito larva*e. G. affinis* feed on *Culex, Anopheles* and *Aedes* larvae of mosquitoes very effectively. This fish is can develop as an adult within four months. It consumes about 165 larvae within 12hrs. Thus, it has great potential in suppression of mosquito larvae populations.

Utility

It is very famous as biocontrol agent for mosquitoes. It also feed on aquatic and terrestrial insects that fall in the water.In natural habitats it is effective against *An. subpictus* species of mosquito causing about 66 per cent mortalities in larvae by feeding upon them. It is equally effective against *Ae. aegyptii*. Thus, *G.affinis* controls Malaria and Dengue and Chikungunya by reducing mosquito populations. In man made habitat like well *An. stephansi* can be controlled with good results within 2 years. In Casurina pits *An. subpictus* has been controlled successfully in Pondicherry. In a costal village of Pondicherry*An. subpictus* is controlled with 96 per cent population while, in rice field *An. freeborni* and *An. pulcherrimus* have been controlled by releasing *G.affenis* with 250 to 750 densities per hectare. It is also extraordinarly effective against the mosquitoes *An. culicifacies* and *An. subpictus* in paddy fields.

2. Gold Fish

Order–Cypriniformes

Family–Cyprinidae

Genus–*Carassius*

Species–*auratus* (Linnaeus) 1758

Distribution–China, Korea, Taiwan, Japan, Europe, Siberia, East Asia, Kampuchea etc. In India, it is introduced as an aquarium fish.

Marks of Identification

1. It measures about 200-460mm in body length.
2. It is beautiful orange colour fish with its slightly bifurcated caudal fin.
3. Dorsal fin covering most of part of dorsal of mid region of the tail.
4. Small anal and ventral fin present. It is with white shining body, slightly laterally compressed.

Special Features

It has been imported to various countries as an ornamental fish however, it has tremendous potential in mosquito larvae control as it easily feed upon mosquito larvae. It has high cost hence, not widely used in mosquito control programmes.

Utility

Used as biocontrol agents for mosquito larvae, specially, *Aedes* and *Anopheles* species. It is effective against *Cx. quinquefasciatus* and *Ar. subalbatus* in containers while, in field condition, *An. subpictus* is controlled upto 34 per cent by introduction of *C. auratus*.

3. Guppy Fish

Order–Cyprinodontiformes

Family–Poeciliidae

Genus–*Poecilia*

Species–*reticulata* (Peters) 1859

Distribution–It is native of tropical America and distributed in the Netherlands, West Indies, Western Venezuela to Ciuvana. It was imported to India more than once. In India it is reported from south India, Maharashtra and some other parts.

Marks of Identification

1. Its male measures about 2 cm and female 4 cm in body length.
2. Body is elongated.
3. Mouth is upturned and head is small.
4. It is whitish dark with dorsal fin touching nearly base of caudal fin, there is colour variation.
5. Caudal fin typically fan like and large with dark and white patches.

Special Features

P. reticulata survives on artificial diet and prefers mosquito larvae for feeding. It can tolerate pollution more than Gambusia in natural habitat.

Utility

It is widely used as biocontrol agent for mosquito larvae. It feed on the larvae of *Culex, Anopheles* and *Aedes* mosquito very effectively. It is supposed to be ideal fish for control of mosquitoes in India and used in mosquito control programme in various states of India including Maharashtra, Karnataka, Kerala, Tamil Nadu, etc. In natural habitat it is effective against *An.subpictus*. In Rice fields it is found effective against *An.aconitus* in Java. In man made habitats it controls *An.gambiae* (in cisterns) *An.subpictus* and *An. stephensi* (in containers) and *Cx.quinquefasciatus* (in Drains), while, in wells *P.reticulata* controls *An.fluviatilis* and *An.culicifacies*.

4. Killi Fish

Order–Cyprinodontiformes

Family–Nothobranchiidae

Genus–*Nothobranchhis*

Species–*guentheri* (Pfeffcr) 1859

Distribution–East Africa: Mombassa to the Pangani River in Tanzania.

Marks of Identification

1. It measures about 7 cm and females are smaller than males.
2. It is with red tail fin, dorsal fin with white and red coloured and pelvic fins are yellow.
3. Behind the eyes four red oblique lines present.
4. Head dorsally with yellowish orange colour with bluish ting, eyes are black ringed with white shining marking. Dorsal fins are spiny.
5. Two males found struggled in a rearing spot.

Special Features

N. guentheri is a fast growing fish. It can grow within 4 weeks as spawning adult. Female lays about 20-100 eggs per day for the whole life.The fish survives until the pool dries up during the dry season.

Utility

N. guentheri was the most suitable fish for antimalaria programme. It is most suitably used as biocontrol agent for mosquito larvae. It can feed on the larvae of *Anopheles, Culex,* and *Aedes* mosquitoes. It has special preference to *Anopheles* spp.

5. Freshwater Gar Fish

Order–Beloniformes

Family–Belonidae

Genus–*Xenentodon*

Species–*cancila* (Hamilton-Buchanan)1822

Distribution–India, Pakistan, Bangladesh, Sri Lanka, Myanmar and Thailand.

Marks of Identification

1. It measures about 30-40 cm in body length.

2. It is niddle like fish, elongated with a long beak like jaws filled with teeth.
3. Dorsal and anal fins are found far back along the body and close to tail.
4. Fish is silvery green darker above and lighter below with a dark band running horizontally along the flank.
5. Males having anal and dorsal fins with a black edge.

Special Features

This is an elegant surface living fish. In North Bengal, it is found in clear, gravelly, perennial streams and ponds of Terai and Duars. It is fairly common in the Ganga-Brahmaputra system of india.

Utility

Used as biocontrol agent for mosquito larvae. It can feed upon the larvae of *Anopheles* and *Culex* mosquitoes. *X. cancila* is effective against *An. subpictus*, *Cx. quinquefaciatus* and *Ar. subalbatus*.

6. Mozambique Ciclilid, Tilapia

Order–Beloniformes

Family–Belonidae

Genus–*Oreochromis*

Species–*mossambica* (Peters) 1852

Distribution–East Africa. It is introduced species in India, Pakistan and Sri Lanka.

Marks of Identification

1. It is large size, 32 cm long and 3 kg in wt.
2. It is laterally compressed with deep and long dorsal fins.
3. Dorsal fin is with spines, it has a weak banding.
4. It is with yellowish body colouration although unreliable with the colouration.

Special Features

O. mossambica grows fast and attains a maximum large size and reproduces under salinities as high as 35 per cent. The lower lethal temperature for this species is 10°C.

Utility

It is effective biocontrol agent of mosquitoes. It feed on *Culex* and *Anopheles* mosquito larvae.

7. Nile Tilapia

Order–Beloniformes

Family–Belonidae

Genus–*Oreochromis*

Species–*niloticus niloticus* (Linnaeus)

Distribution–East Africa. West Africa, River Nile.

Marks of Identification

1. It measures about 34 cm in length.
2. It is laterally compressed with large single dorsal fin with spine.
3. Paired pelvic fin, single ventral fin and single anal fin.
4. Mouth blunt, notched, operculum large sized.
5. Banded laterally with black and grey marking.
6. Dorsal head region blackish brown.

Special Features

O. niloticus niloticus is the fastest growing species in many countries. It does not tolerate high salinity.It is also poor cold tolerant but, highly suitable for farming in tropicalclimate, fresh water and brackish water systems. The lethal temperature is 12°C.

Utility

It is potential biocontrol agent of mosquito larvae in fresh water for *Aedes* species and brakish water for *Anopheles* and *Culex* species.

Discussion and Conclusion

Fishes are very good source of mosquito control. Therefore, they are playing an increasing role in vector management strategies of many mosquito borne diseases. However, biocontrol technology is challenging as well as difficult. Unlike the chemical pesticides, the results are often unpredictable with biological agents. Therefore, for a better understanding of biocontrol programme, the biological interactions with the environment are very much essential to understand. Thus, developing and acquiring the necessary skill assume paramount importance. In India, the success of such mosquito control strategies depends on developing simple technology backed by a campaign of public education. Targeting vectors of diseases is the most effective strategies to control vector-borne diseases. Many recent efforts made by workers in new mosquito control tools are confined to the laboratory only. However, biological agents carry the potential for overcoming such obstacles.Therefore, there is need of programme of biological research aimed towards understanding the factors that limit the number of mosquitoes through biological means. The research should be installed towards testing efficacy of biocontrol agents of mosquitoes. The search efficiency of the introduced predator and prey selectivity patterns of larvivorous organisms should be exploited properly by offering mosquito larvae in combination with other alternate natural preys. The efficacy of the biocontrol agents are depends on a multitude of factors such as - (1) Characterization of natural enemies including taxonomical, morphological, ecological, or genetic markers. (2) Selection of climatically suitable biocontrol agents.(3)Evaluation of semi-field or field cage conditions (4) quarantine evaluations of BCAs prior to proceeding with natural release (5) assessment of unintended impacts and (6)assessment of efficacy of existing indigenous biocontrol agents.

As regards to the fish biocontrol agents of mosquitoes, therefore two main factors determines the efficacy and suitability which includes the ability of the fish to eat enough larvae of vector species to reduce the number and the suitability of the breeding environment of the mosquitoes. The second factor is best known by finding a native fishes that thrives under the conditions prevalent in breeding sites rather than to change breeding sites to accustom the fish. According to Wu *et al.* (1991) a ditch ridge system for rice fields is essential for better accommodation of fish. However, it is kept in mind that the use of pesticides and fertilizers have negative influence on fish stock in irrigated fields. The first factor may be strongly influenced by aquatic vegetation, which in turn, can interfere with fish feeding and can also provide refuge for the mosquito larvae. The effectiveness of fishes to control mosquitoes is vary and dependant on environmental complexity. Therefore, efficacy of several indigenous larvivorous fishes in different season in wetlands and larger water bodies should be evaluated. Their role in the tropic cascade from the community view point has also tremendous importance.

It is very much essential to know the disadvantages of using larvivorous fishes. *Gambusia* when stocked in waters outside their native range, it cause serious negative ecological impacts. Because, *Gambusia* is an opportunistic predator.It has highly variable diet which feed on algae, zooplankton, aquatic insects, as well as eggs and young ones of fish and amphibians. According to Gracia-Berthou (1999) *Gambusia* shows a diet shift from diatoms to cladocerans to adult insect with the maturation.It is voracious and highly aggressive fish which compete with the native fishes very successfully for viable food and space. *Gambusia* decreases all large zooplanktons but, rotifers and phytoplankton densities are found increased.It also consumes a high percentage of the phytoplankton grazers. *Gambusia* fishes indirectly cause adverse ecological changes such as increased phytoplankton abundance, higher water temperatures, more dissolved organic phosphorous and decreased water clarity. Therefore, periodic removal of vegetation is essential to facilitate the activity of the fishes. Finally, rather in temporary or ephemeral mosquito habitats, other forms of biocontrol agents must be protected and exploited carefully without causing harm to nature.

References

Chandra, S.K. and Deuti, K., 1997. Endemic amphibians of India. *Rec. Zool. Surv. India*, 96(1–4), 63–79.

Chatterjee S.N., Chandra G., 1996. Laboratory trails on the feeding pattern of *A. subpictus, C.quinquefasciataus* and *Armigera subalbatus* by *Xenentodon cancila* Fry. *Env. Ecol.*, 14: 173–174.

Chatterjee S.N., Das S. and Chandra, G., 1997. Gold fish as a strong larval predator of mosquito. *Trans. Zool. Soc., India*, 1: 112–114.

Daniel, J.C., 1963. Field guide to the Amphibians of Western India. Part–I. *J. Bombay Nat. Hist. Soc.*, 60(2): 415–438; 60(3): 690–702.

Daniel, J.C., 1975. Field guide to the Amphibians of Western India. Part–III. *J. Bombay Nat. Hist. Soc.*, 72(2): 506–523.

Daniel, J.C. and Sekar, A.G., 1989. Field guide to the Amphibians of Western India. Part–IV. *J. Bombay Nat. Hist. Soc.*, 86(2): 194–202.

Dua, V.K. and Sharma, S.K., 1994. Use of guppy and gambusia fishes for control of mosquito breeding at BHEL industrial complex, Hardwar (U.P). Malarial Research Centre, New Delhi, p. 35–42.

Dutta, S.K., 1992. Amphibians of India, updated species list with distribution record. *Hamadryad*, 17: 1–13.

Dutta, S.K., 1997. *Amphibians of India and Sri Lanka: Checklist and Bibliography*. Odyssey Publ. House, Bhubaneshwar, 342 pp.

Ghosh, A., Mondal, S., Bhattacharjee, I. and Chandra, G., 2005. Biological control of vector mosquitoes by some common exotic fish predators. *Turk. J. Biol.*, 29: 167–171.

Hackett, L.W., 1937. *Malaria in Europe: An Ecological Study*. Oxford University Press, London.

Kim, H.C. and Kim, M.S., 1994. Biological control of vector mosquitoes by the use of fish predators, *Moroco oxycephalus* and *Misgurnus anguillicandatus* in the laboratory and semi field rice paddy. *Korean J. Entomol.*, 24: 269–284.

Kramer, V.L., 1989. The ecology and biological control of mosquitoes in California wild and white rice fields. Desseration abstracts international. *Bull. Sci. Engin.*, 49: 4670B.

Martinez, J.A. and Guillen, Y.G., 2002. Indigenous fish species for the control of *Aedes aegypti* in water storage tanks in southern Mexico. *Bio Control*, 47: 481–486.

Milam, C.D., Farris, J.L. and Wilhide, J.D., 200. Evaluating mosquito control pesticides for effect on target and non-target organisms. *Arch. Environ. Contam. Toxicol.*, 39: 321–328.

Prasad, H., Prasad, R.N. and Haq, S., 1993. Role of biological agents for the control of mosquito breeding through *G. affinis* in rice fields. *Indian J. Malarial.*, 30: 57–65.

Rajnikant, Pandy, S.D. and Sharma, S.K., 1996. Role of biological agents for the control of mosquito breeding in rice fields. *Indian J. Malarial.*, 33: 209–215.

Sathe, T.V., 1996. Role of Amphibians in insect pest control. *Proc. Biol. Control. Insect. Pests*. Workshop CSIR and DST. P.11.

Sathe, T.V., 2006. Biological control of mosquitoes. *Proc. Nat. Sym. Rec. Trend Malarial Studies*, pp. 11–17.

Sathe, T.V. and Bhoje, P.M., 2005. Biocontrol potential of Guppy *Poecilia reticulata* (Peters) for mosquitoes. *J. Expt. Zool., India*, 8(1): 105–108.

Sathe, T.V. and Gire, B.E., 2002. *Mosquitoes and Diseases*. DPH, Delhi, pp. 1–121.

Sathe, T.V., Jagtap, M.B. and Sathe, Asawari, 2010. *Mosquito Borne Diseases*. Mangalam Publ., New Delhi, pp. 1–335.

Saha, D., Biswas, D., Chatterjee, R.C. and Bhattachary, A., 1986. Guppy as a natural predator of *Culex quinquefasciatus* larvae. *Bull. Cal. Sch. Med.*, 34 : 1–4.

Sergeev, B.V., 1986. *The World of the Amphibians*. Mir Publishers, Moscow, pp. 188.

Service, M.W., 1983. Biological control of mosquitoes–has it a life. *Mosq. News*, 43: 113.

Sharma, V.P. and Ghosh, A., 1989. *Larvivorous Fishes of Island Ecosystem*. Malarial Research Certre (ICMR), New Delhi.

Waage, J.K. and Greathead, D.J., 1988. Biological control, challenge and opportunities. *Phil. Tras. R. Soc., Lond.*, 318: 111–128.

Yu, H.S. and Lee, J.H., 1989. Biological control of malaria vector by combined use of larvivorous fish and herbivorous hybrid in rice paddies of Korea. *Korean J. App. Entomol.*, 28: 229–236.

Chapter 22

AMPHIBIANS FOR INSECT PEST MANAGEMENT

T.V. Sathe[1] and P.M. Bhoje[2]

[1]*Department of Zoology, Shivaji University,*
Kolhapur – 416 004, Maharashtra, India
[2]*Department of Zoology, Y.C. Mahavidhyalaya,*
Warnanagar, Kolhapur, Maharashtra, India

Introduction

Amphibians are specialized for living in two types of habitats *viz.* water and land. Amphibians are well suited for living in water but not suited on land since their skin does not have scales or hairs and is not keratized. They can not regulate their internal body temperature therefore, on the land they have to face more difficulties than water.

During the monsoon period, in rainy days males start singing for the females for mating purpose. Several males become available for mating but only one get success to mate and others are kicked out by hind legs. A very large number of eggs are laid by female and females are larger than males mostly twice the size of males. Eggs laid by female are fertilized by males. The pigmented eggs found floating and enclosed by jelly like coating of flat mucilaginous masses/gelatinous masses. In case of tree frogs, eggs are laid in a frothy mass and found hanging to substrates over pool of water. In such cases, naturally, after hatching eggs tadpoles drop down in the water and get appropriate environment for further development. The tadpole do not have the mouth hence they obtain the nutrition from attached sacs. Later, mouth is developed and then start feeding on algae, plankton or carrion by griming. By the time fins are developed for swimming. The tadpoles are then transformed into froglets. Then the froglets move to the land and start feeding on insects. Due to difficult and unsuited life on land it is supposed that the only 2 per cent froglets become adults and rest dies. Secondly, there is tremendous pressure of pesticides, sewage, industrial toxic effluents on animal and plant wealth including human being and the amphibians. Their use in insect pest management will definitely prove their worth. Many species of amphibians are disappearing due to transformation of natural habitat to land for cultivation and urbanization. Therefore, there is need to protect and conserve the Amphibians. Amphibians have very important role in natural food web. They circulate organic and inorganic between land and water. Amphibians largely subsists upon insect diet and are responsible for suppressing insect pest populations and disease vectors. In past, some workers (Daniel, 1963, 1965; Dutta, 1992; Mohanty and Acharya, 1982; Sarkar, 1984; Sathe *et al.,* 2006) attempted biodiversity of amphibians and their role in insect pest management.

From India, 219 species of amphibians have been reported (Dass and Dutta, 1998). Review of literature indicates that highest concentration of species was found in Western peninsula. Out of the 219 amphibians species of India, 134 species are endemic to the country. From Western peninsula 92 species are reported to be endemic and 9 are critically endangered. The status of most of the Indian amphibians is unknown to the science. Frog limbs export business is gaining importance day by day in India. Therefore, frogs are killed on large scale due to which Govt. of India banned export of frog legs since April, 1986. Hence, any advance knowledge on this animal group will add tremendous importance in saving and utilizing the diversity of Amphibians for sustainable development of our country.

1. Marbled Toad (*Bufo stomaticus* Litken, 1862)

Distribution

India, Pakistan, Afganisthan, Nepal, SriLanka and Myanmar. In India it has been reported from Maharashtra, Karnataka, Tamil Nadu, Assam, Kerala and Andhra Pradesh etc.

Marks of Identification

It is medium sized with head broader than long. Its snout is rounded. Toes 2/3 webbed with 2 segments of 4th one are free. Parotoid glands are large and not bean shaped but flat and elliptical, upper body part is brownish or olive grey. Belly with dull whitish. Light brown coloured juveniles are with dark marblings. Males are smaller than females and with bright yellow faint and black vocal sacs. It is more active than Indian toads.

Life Cycle

It has breeding period from June to September. In monsoon male gives call to female. Copulation lasts for 1-4 days. Competition for mate in males is present in this species. Oviposition started at the onset of monsoon. In two parallel strings pale yellowish green eggs are laid. A single mated female can lay about 10,000 eggs. They hatch within 24 hours. The tadoles are about 12 mm in body length with moderately flat head and body and narrowly oval. It has rounded snout. Black coloured body of tadpole contain shiny white silvery spots.

Non-Insect Preys

Earthworms, centipedes, molluscs and spiders.

Insect Preys

Its food largely consists of various kinds of insects such as Termites, crickets, grasshoppers, mole crickets, caterpillars, earwigs, ants, bees, moths, cicadas, plant bugs, jassids, delphacids, aphids, mealy bugs, ground beetles, bark beetles, click beetles, weevils, may flies, mosquitoes etc.

Utility

It acts as biocontrol agent of above insect pests.

2. Cane Frog (*Bufo marinus*)

Distribution

America, various Islands, Australia, Pacific and Caribbean regions, Jamaica, Puerto Rico, Barbados, New Guinea, Philippines, Japan, Sweeden etc.

Marks of Identification

It measures from 10-15 cm except appendages. It may even measures 38 cm to 54 cm in certain countries. They can survive for 10 to 15 years in open environment. It is dry and warty skinned frog with distinct ridges above. The eyes running down to snout, parotoid gland is large and found behind eye. Horizontal pupils have golden irise. Fingers without webbing.

Life Cycle

It breeds in freshwater. After mating female lay black eggs on water and covered with jelly like substance. A single female can lay about 10,000 to 25,000 eggs. Incubation period is 48 hours (range 14 hours to 7 days). The tadpoles are small, black and with short tails. The development from egg to toadlet required 12 to 60 days. Very interestingly, eggs and tadpoles are toxic to several organisms. In adult-hood stage *B. martinus* produce *"bufotoxin"*.

Non-Insect Preys

Preys are detected by this species with the help of sense of smell. They predates upon small rodents, reptiles, amphibians, birds and several non insect invertebrates.

Insect Preys

Almost all kind of insects are taken by this species as food. The more important insects refer to cockroaches, caterpillars, grasshoppers, crickets, mole crickets, termites, jassids, cicadas, plant bugs, termites, ants, wasps, beetles, weevils, mealy bugs, dragonflies, butterflies, moths etc.

Utility

It is mass reared in Australia and used as biocontrol agent against several insect pests. In Australia froggery of *B. marinus* is very important trade.

3. Green Pod Frog (*Rana hexadactyla* Lesson 1834) (Order - Anura)

Distribution

SriLanka, Bangladesh, India. In India it is found in Maharashtra, Andhra Pradesh, Karnataka, Tamil Nadu, Orissa, West Bengal, Gujarat etc.

Marks of Identification

It measures from 50-140 mm in length (snout to vent). It is called giant leaf green frog. Its head is as long as broad. Snout is rounded and flat and slightly pointed. Tip of fingures and toes are pointed. A strong dermal fringe is present on the outer toes. Toes are fully webbed. Outer tubercle absent but inner tubercle is elonged and pointed. There are two rows of porous warts along the flanks and lower surface is more or less granular. Body is olive green or bright grass green dorsally. Throat colour is yellow or stripped with brown. Juveniles are with black or dark green spots and bars on the back. Males are with external vocal sacs. It is found in floating in ponds and fishing ferries along with *Eichorrnia, Azolla, Pistia* etc. and very active, can lead for longer distance.

Life Cycle

Its breeding period starts from June-July and extends for October. A female can lay about 2000 eggs in ponds or in paddy fields. Tadpole is with more produced heavy black snout. It is smaller and more slender than *R. cynophylctis* and with narrow but longer tail. Dorsally it is olive green with dark blotches and ventrally whitish. Dark spots are absent on the tail. Silvery spots present on the side of the head, body and tail. Its snout to vent length is about 25 mm.

Non-Insect Preys

Snails, small fishes, even small frogs. This species largely subsists upon weeds but above preys are also taken.

Insect Preys

Termites, grasshoppers, crickets, moths, jassids etc. Adults feed on mosquitoes, dragonflies and tadpoles feed on naids (Larvae) of dragonflies and mosquito larvae in aquatic medium.

Utility

It is useful biocontrol agent of various insects pests mentioned above and has importance as laboratory animal for study purpose. *Bufo marinus* is mass reared in Australia for biological of insect pests. Therefore, this species has also great potential.

4. Indian Bull Frog (*Rana tigerina* Daudin, 1802) = *Limnonectes tigerinus* (Daudin, 1802)

Distribution

China, Thailand, Myanmar, Taiwan, Bangladesh, Nepal, Sri Lanka, Pakistan and India. In India, it is reported from all parts including base of Himalayas.

Marks of Identification

It measures 50-150 mm from snout to vent. Its head is as long as wide. These large sized frogs have smooth skin and longitudinal glandular folds on the dorsal side. Ear-drum is about equal to eye diameter and fingers are without webs. Finger tips and toes are not sharply pointed. An outer fringe of web is present to fifth toe. The frog is with short and blunt inner pedal tubercle. Lower side is smooth and porous warts absent on the flanks. This species have yellowish or olive green or brownish green colour from dorsal side and leopard like dark spots. Dorsal also contain a yellowish median stripe running from snout to vent. Males are smaller and darker than females and with bright blue coloured vocal sacs. It is semiaquatic swimmer. Mostly live solitarily except breeding season. They may aestivate during dry season.

Life Cycle

After onset of monsoon/rainy season they proceed with aestivation for breeding. Breeding period is from June to September. Males give the call for females on either in paddy field water or rain water pools, or ditches. Several males gather towards a single female but, only one succeeds in holding the female. After copulation eggs are laid in temporary rain water pools. From which tadpoles are given out. These are omnivorus and mostly bottom feeders. The tadpole is characterized by having black colour entirely and tail about twice the length of body. The tadpoles are finally transported into the froglets and then matured frogs.

Non-Insect Preys

This species has very wide range of animals for food. It feed on land crabs, earthworms, snails, centipedes, scorpions, spiders, small fishes, lizards, small water snakes, small birds, mice, squirrels etc. This species is cannibalistic in nature.

Insect Preys

Tadpoles feed on mosquito larvae while, adults feed on several insect pests such as Lepidopterous caterpillars, grasshoppers, cockroaches, termites, crickets etc.

Utility

Tadpoles are utilized as biocontrol agent for mosquito larvae in aquatic medium while, adults acts as very good biocontrol agent for above mentioned insect pests.

5. Warty Frog (*Rana keralensis* Dubois)

Distribution

India. In India, it has been recorded from Maharashtra, Karnataka, Tamil Nadu, Kerala, West Bengal and Gujarat. In Maharashtra it is found in plains.

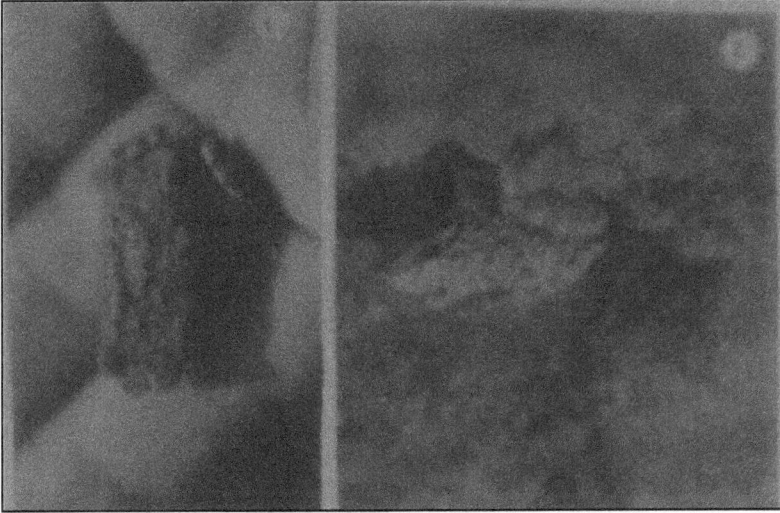

Figure 22.1: Toad Sand.

Figure 22.2: Tree Frog.

Marks of Identification

It measures about 40-70 mm from snout to vent and medium sized. Head is slightly broader than long. Snout is about pointed. It has large eyes with dilatable and rounded pupil. Ear drum is prominent and nearly equal to the diameter of eyes. Fingure tips are pointed. Toes almost webbed with two segments of 4th toe free. Eyes are black. Back skin is rough and warty and skin on belly is smooth. It is cryptic coloured with variation. It has camouflaging capacity against soil and rock background. Small dark markings present on upper part of dark grey or brown

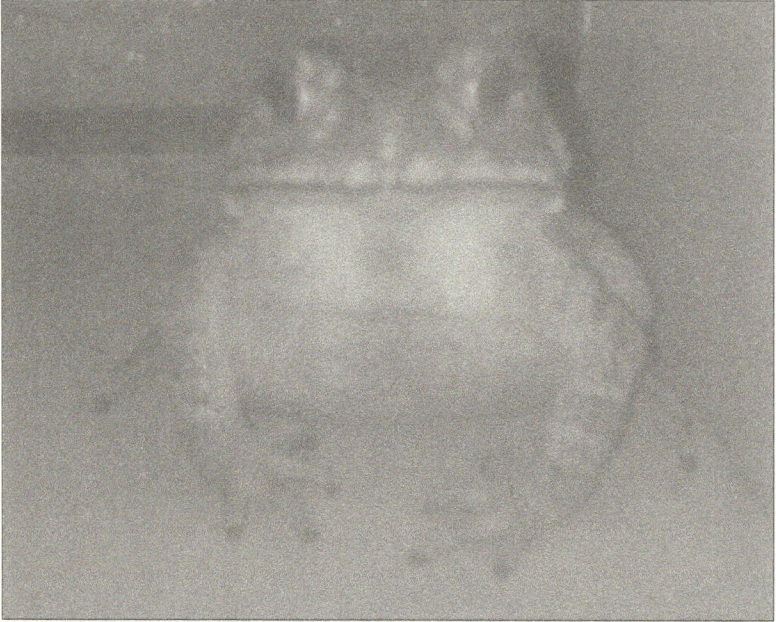

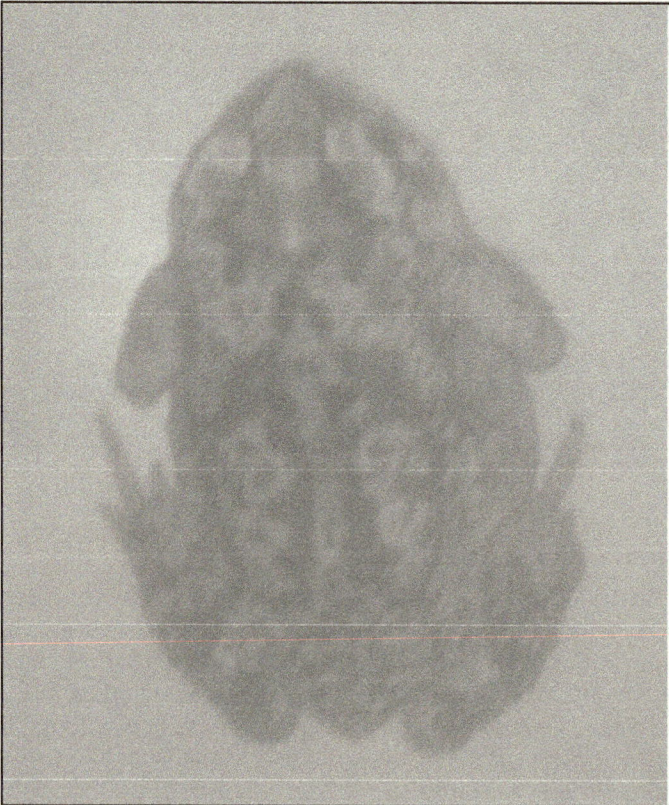

Figure 22.3: Toad *Bufo* sp.

colouration. In males internal vocal sacs are present. Adults are found in thick grasses and bushes, streams in forests and are mostly nocturnal. It is short leaped frog.

Life Cycle

There is competition and aggressive behaviour for mate between males. The male gives the call for mating with a series of 9-11 croaks. Mustard seed like eggs are laid in shallow water pools. They are laid in batches of 200-300 at 5-16 days interval. Incubation period is 30 hours. Eggs hatched into tadpoles. The tadpoles are brown, elongated and found within jelly. After 18 hours they become free swimmers. Later, they become transparent. On their body dark brown pigmented spots are found. Initially tadpole is herbivorus and bottom feeder. Then it becomes carnivorous. Within 45 days hind legs are developed to tadpoles and then after 15 days 10 mm long froglets are developed and then disperse to wet soil and litter for further life.

Non-Insect Preys

Earthworms, their own tadpoles (cannibalism is present in older tadpoles) small, animals from aquatic invertebrates.

Insect Preys

Caterpillars, termites, crickets, cockroaches, mosquitoes, grasshoppers, ants, moths, small beetles etc.

Utility

It acts as very good biocontrol agent for above mentioned insect pests.

6. Burrowing Frog [*Tomopterna breviceps* (Schneider, 1799)]

Distribution

Myanmar, Bangladesh, Pakistan, India, Nepal and SriLanka. From India, it has been reported from Maharashtra, Karnataka, Andhra Pradesh, Tamil Nadu, Orissa, West Bengal, Bihar, Uttar Pradesh, Rajasthan etc.

Marks of Identification

It measures for 20-50 mm from snout to vent in length. Its head is broader than long and snout is short and rounded. Ear-drum is more or less half the diameter of the eye. Its fingures are without webs but with swollen tips and not disc like. Back skin is smooth but belly is with granular skin. Nostril is at equal distant from the tip of snout and eye. Dorsal of the frog is yellowish brown or grayish with dark spots or markings. Lower side is pinkish white. Juveniles are pale brown with dark spots. Male shows paired black vocal sacs. This is nocturnal and terrestrial toad.

Life Cycle

After heavy monsoon males create in pairs near pools of water or gardens. Breeding period is from June to August. The frog enter in the water only in breeding period. Eggs are laid on water surface as cluch floating. Incubation period is about 45 hours. Tadpole shows yellowish brown to deep grey colour dorsally, further, it contain brown spots and marking on the back. It also shows M-shaped mark on the back. In between the eyes there is 'V' mark. The above marks help in differtiating tadpole of this species from others. Within 18-20 days tadpoles developed into froglets. The tadpoles are always bottom feeders.

Non-Insect Preys

Spiders.

Insect Preys

Termites, crickets, cockroaches, wasps, ants, grasshoppers, beetles, etc.

Utility

It acts as very good biocontrol agent for above mentioned pests.

7. Balloon Frog [*Uperodon globulosus* (Gunther, 1864)]

Distribution

India, Bangladesh. In India, it has been reported from Maharashtra, Karnataka, Madhya Pradesh, Bihar, Orissa and Gujarat.

Marks of Identification

It measures 40-66 mm from snout to vent. The small head is more broader than long and sout is rounded. Its tongue is oval. Tips of figures are rounded and not bearing discs. It has short hind legs. It shows shovel shaped two pedal tubercles. Dorsally skin is smooth but, wrinkled below. Its back colour is reddish brown to greenish grey or dirty white from undersurface. Males are darker. It is burrower and prefer loose sandy soil. It dig out termataria for feeding on termite eggs and other castes of termites.

Life Cycle

Breeding of this species starts with initiation of rainy season from June to July. Male gives loud croaking call to females for mating. During mating the pair remain tightly fixed due to certain secretion. The pair enjoy several dips in the water at the end of mate and finally get separated. The eggs laid are coated with jelly-like substance. Eggs increase in size due to hydration. The eggs are hatched into active tadpoles. They have olive brown colour from above and spots from under side specially on flanks. Tadpoles are further characterized by having whitish tail with blochy and longitudinal stripes.

Non-Insect Preys

Not known.

Insect Preys

It feed voraciously on winged and non winged termite castes by digging termataria. The frog also predates upon small beetles, caterpillars, grasshoppers, jassids etc.

Utility

U. globulosus is very good biocontrol agent of above mentioned insect pests and helps in keeping environment ecofriendly by reducing risk of pesticides.

8. Malbar Gliding Frog (*Recifos malbaricus*)

Distribution

India: In India, it has been reported only from Western Ghats. Maharashtra specially from Amboli, Sindhudurg district.

Marks of Identification

It is parrot green coloured from above and light coloured from below. The legs are with red coloured webs. Its eyes are white with horizontal black pupil is the beauty of this species. It is found on trees jumping from one branch to another. This species is found only at Amboli Ghats of Maharashtra.

Life Cycle

Its life cycle is not fully illustrated. However, egg, tadpole, froglet and matured frog are different stages of this species. Eggs are laid near in water. Tadpoles are aquatic and froglets are terrestrial and matured frogs are arborial or they are tree frogs.

Non-Insect Preys

Not known.

Insect Preys

Termites, grasshoppers, moths, butterflies, Jassids, mealy bugs, scales, aphids, white flies, caterpillars etc.

Utility

It can be used as biocontrol agent for insect pests of forest trees. Several other amphibians have greater role to play in biological control of insect pests. Therefore, their check list, biology and utility should be studied on priority basis.

Distribution pattern of Amphibians from Western peninsula is given below (Table 22.1).

Table 22.1: Distribution Pattern of Amphibians in W. Peninsula.

Region	States Included	Orders	Genera	Species
Western peninsula	Kerala, Karnataka, T.N.(West), Goa, Western Maharashtra	Anura	19	110
		Gymnophiona	4	22

Fauna of Amphibians predating on insect pests of forest trees from Western Ghats is given in Table 22.2.

Table 22.2: Amphibians Predating on Forest Insect Pests from Western Ghats of Maharashtra.

Sl.No.	Species	Orders	Insect Preys
1.	*Ansona ornate* Gunther	Anura	Termites, grasshoppers
2.	*A. rubigina* P. and P.	Anura	Grasshoppers, crickets
3.	*Bufo beddomis* Gunther	Anura	Termites, moths
4.	*B. brevirostris* Rao	Anura	Moths, Jassids
5.	*B. hololius* Gunther	Anura	Termites, moths, beetles
6.	*B. koynayensis* Soman	Anura	Dragonflies, termites
7.	*B. parietalis* Boul.	Anura	Termites, beetles, ants, Jassids
8.	*B. silentvalleyensis* Pillai	Anura	Flying insects
9.	*Rana aurantiaca* Boul.	Anura	Grasshoppers, Ants
10.	*Rana malbarica* T.	Anura	Beetles, moths
11.	*R. curtipes* Jerdon	Anura	Spider, mosquitoes
12.	*R. temporalis* Gunther	Anura	Mole crickets, termites
13.	*Philautus nerostagona* B.&B.	Anura	Mosquitoes, spiders
14.	*P. signatus* Boul.	Anura	Crickets, grasshoppers
15.	*P. travancoricus* Boul.	Anura	Mosquitoes, moths
16.	*P. tuberohumerus* K. and J.	Anura	Termites, moths,
17.	*Nyctibatrachus aliciae* Inger	Anura	Moths, mosquitoes
18.	*N. deccanensis* Dubois	Anura	Moths, termites, Ants
19.	*N. major* Boul.	Anura	Grasshoppers, moths, beetles
20.	*Polypedates pseudocrucigar* D. and R.	Anura	Termites, cockroach, ants, grasshoppers, etc.
21.	*Rhacophorus malbaricus* Jerdon	Anura	Moths, mosquitoes
22.	*R. pseudomalbaricus* V. and D.	Anura	Cockroaches, moths, beetles, Ants
23.	*Gegeneophis fulleri* Alcock	Gymnophiona	Grasshoppers, Ants
24.	*G. krishnii* P. and R.	Gymnophiona	Moths, mosquitoes
25.	*Indotyphlus maharashtraensis*	Gymnophiona	Moths
26.	*I. bombayensis* Taylor	Gymnophiona	Moths, termites
27.	*I. malbarensis* Taylor	Gymnophiona	Moths, Ants
28.	*I. peninsularis* Taylor	Gymnophiona	Termites, moths
29.	*Uperodon globulosus* Gunther	Anura	Termites
30.	*Indirana beddomii* Gunther	Anura	Beetles, grasshoppers
31.	*I. diplostica* Gunther	Anura	Termites, crickets, Ants, moths etc.
32.	*I. longicrus* Rao	Anura	Long horned, grasshoppers, moths,
33.	*Philautus jayarami*	Anura	Termites, moths
34 .	*P. akroparallagi*	Anura	Moths, termites
35.	*P. sushili*	Anura	Moths, Ants, termites

Reference

Chandra, S.K. and Deuti, K., 1997. Endemic amphibians of India. *Rec. Zool. Surv. India*, 96(1–4): 63–79.

Daniel, J.C., 1963. Field guide to the Amphibians of Western India. Part–I. *J. Bombay Nat. Hist. Soc.*, 60(2): 415–438; 60(3): 690–702.

Daniel, J.C., 1975. Field guide to the Amphibians of Western India. Part–III. *J. Bombay Nat. Hist. Soc.*, 72(2): 506–523.

Daniel, J.C. and Sekar, A.G., 1989. Field guide to the Amphibians of Western India. Part–IV. *J. Bombay Nat. Hist. Soc.*, 86(2): 194–202.

Dutta, S.K., 1992. Amphibians of India, updated species list with distribution record. *Hamadryad*, 17: 1–13.

Dutta, S.K., 1997. *Amphibians of India and Sri Lanka: Checklist and Bibliography.* Odyssey Publ. House, Bhubaneshwar, 342 pp.

Sathe, T.V., 1996. Role of Amphibians in insect pest control. *Proc. Biol. Control Insect Pests.* Workshop CSIR and DST. P.11.

Sergeev, B.V., 1986. *The World of the Amphibians.* Mir Publishers, Moscow, pp. 188.

Chapter 23

BIRDS IN INSECT PEST MANAGEMENT

T.V. Sathe, P. Ranbhare Minakshi
and A. Valivade Rupali*

Department of Zoology, Shivaji University,
Kolhapur – 416 004, Maharashtra, India

Introduction

Birds are very important part of our natural heritage. They first appeared in 150 million years ago. There are about 10,000 species (30 per cent) of world listed. They are reported from highest Himalaya to great Mangroove forests. In India, more than 45 species of birds are recognized as endemic and 26 are very rare or accidentally found.

With more than 1250 species of birds, India and the rest of South Asia are paradise for bird watching. There are over 925 breeding species in India. The Indian birds belong to 75 families and 20 orders. They found in different regions of our country *viz.* Trans Himalyayan, Eastern Ghats, North east coasts and Nicobar Islands. The Blue peafowl is the national bird of India. Birds are economic wealth of India from the view point of nutrition of increasing human population, maintaining ecological processes, pollination and biological control of pest insects as biocontrol agents. Our Indian Mynah bird *Acridotheres tristis* has been used in first International movement of biological pest management in 1772 in Moritious against red locust *Nomadacris septumpunctata*, pest of sugarcane.

* E-mail: profdrtvsathe@rediffmail.com

Chemical control never solve the permanent problem of pest control and lead serious problems such as air, water, soil pollutions, health hazards, killing of beneficial organisms, pest resistence, secondary pest outbreak, pest resurgence, interruption to ecocycles and food webs, etc. The above things clearly indicate that there is need to have an alternative for use of chemical pesticides. On this background birds can play important role in ecofriendly control of insect pests. The present work will add great relevance in biological control of insect pests. Hence, the present topic is visualized.

Materials and Methods

Biodiversity of birds have been studied by spot observations, one man one hour search basis from Kolhapur region (Shivaji University and Tahasils from Kolhapur district). The birds have been observed at morning hours from 7.00am. to 9.00pm. and photographed by zoom lens, Cannon camera for their description. Morphological characters have been described and their identification was made by consulting appropriate literature. Distribution of birds have been studied by one man one hour spot observation method from different places of Kolhapur region. Life cycle of birds have been studied by consulting literature and through questionary to local people and by spot observations. Insect preys to birds have been identified by spot observations. Later, the insects have been collected and identified by consulting appropriate literature. The observations of the birds were also made at evening and morning when cattle were grazing and during ploughing of the land in the region.

Results

The insectivorous birds as biocontrol of insect pests from Kolhapur region are described below. Their photographs are shown in the Figures A to T.

1. Indian Mynah (Plate 23.1A)

Taxonomic Position

> Class – Aves
>
> Order – Piciformes
>
> Family – Sturnidae
>
> Genus – Acridotheres
>
> Species – tristis

Distribution

Widely distributed in India, Bangladesh, etc.

Occurrence

Birds are seen near human habitation, lawns, grassy road side, etc. Indian Mynah bird found throughout the year.

Marks of Identification

Indian Mynah is locally called as "Salunkhi". The bird measures about 20.5 cm in size. It has light yellow beak and eyes black ringed with yellow, small feathers, neck is also having black feathers. Ventrally it shows whitish brown feathers. Its legs are yellow. Tail is decorated with black feathers and head is black. A large white patch is located on the wings.

Life Cycle

Breeding season begins in April and continuous till August. Sexes are alike. Mynah's nest is a collection of paper straw, rubbish into hole in a free or in wall or ceiling of buildings, etc. Mated female lays about 4-5 eggs in the nest. The eggs are beautiful glossy blue without any markings. After hatching the eggs, young chicks are fed by parents and grown to wing formation in the nest. Later, they feed individually with the parents.

Insect Preys

Insects like Cockroaches, grasshoppers, crickets, locusts, red locusts and several small insects are eaten by this bird.

Economic Importance

It is biocontrol agent of red locust, very first bird used in International Biological Control Programme of insect pests in Mouritious in 1772. It is very effective biocontrol agent of insects associated with wheat, jowar, pomegranate, sugarcane, etc.

2. Black Drongo or King Crow (Plate 23.1B)

Taxonomic Position

Class –Aves

Order – Piciformes

Family – Dicruridae

Genus – *Dicrurus*

Species – *adsimillis*

Distribution

India, China, Sri Lanka, Iran, Indonesia, etc.

Occurrence

Birds are seen near human habitation, lawns, grassy road side, etc. This bird found throughout the year.

Marks of Identification

The bird is bright black coloured with bifurcated tail. Adults usually have a small white spot at the base of the gape. The iris is dark brown. It is aggressive and fearless bird. The bird measures about 30 cm in body length.

Life Cycle

The female bird can lay about 3-5 eggs. They are whitish with brownish red spots. Eggs are laid close to the first rains. First year birds have white tips to the feathers of the belly. While, second years have these white tipped feathers restricted to the vent. The eggs are hatched in the nest, young ones are provided food by their parents and full grown by wing development.

Insect Preys

Grasshoppers, cicadas, termites, wasps, bees, ants, moths, beetles and dragon flies. It is very potential biocontrol agents of several insect pests.

Economic Importance

The bird acts as biocontrol agent of above mentioned insects.

3. Shikra (Plate 23.1C)

Taxonomic Position

Class –Aves

Order – Falconiformes

Family – Accipitridae

Genus – *Accipiter*

Species – *badius*

Distribution

Throughout South Asia and Sub Saharan Africa.

Occurrence

Shikra is a bird of open woodland including savannah and cultivation.

Marks of Identification

This bird is a small raprot measuring about 26-30cm in size. It shows short broad wings and a long tail, both adapted for fast moving. The normal flight of this species is a characteristic "flap-flap-glide". The adult Shikra is about the size of pigeon which has pale grey or ashy blue gray upper parts and white, finely barred reddish below. Sexes are similar except that females are larger than the males. The young birds are brown above and white spotted with brown below. It has a barred tail.

Life Cycle

The bird can lay 3-4 eggs in the nest. Its eggs are pale bluish white, some times faintly spotted and speckled with grey. Upto wing development young birds are fed by parents in the nest. After wing formation they leave nest and work with parents.

Insect Preys

Locusts, grasshoppers, crickets, termites, beetles, etc.

Economic Importance

The bird acts as biocontrol agent of above insects pests.

4. Peacock (Plate 23.1D)

Taxonomic Position

Class –Aves

Order – Galliformes

Family – Galliformidae

Genus – *Pavo*

Species – *cristatus*

Distribution

India.

Occurrence

It is found throughout the year in fields, gardens, forests, orchards, etc.

Marks of Identification

Female has a mixture of dull green, brown and grey colouration in her plumage. She lacks the long upper tail coverts of the male but has a crest. Females can also display their plumage toward off danger to their young or other female competition.

The male has iridescent blue-green or green coloured plumage. The so called 'tail' of the peacock, also termed the 'train' is not the tail quill feathers but highly elongated upper tail coverts. The train feathers have services of eyes that are best seen when the tail is fanned. Both have a head crest.

Life Cycle

Normally 3-5 eggs are laid by mated female. Eggs are pale cream to café-all-lait in colour. As in other birds parents take the care of young one upto the dependable age.

Insect Preys

It feeds on grasshoppers, termites, caterpillars, ants, grubs, crickets, etc.

Economic Importance

The bird acts as biocontrol agents of above insects pests.

5. Red Vented Bulbul (Plate 23.1E)

Taxonomic Position

Class – Aves

Order – Piciformes

Family – Pycnotidae

Genus – *Pycnonotus*

Species – *cafer*

Distribution

Native range covers the Indian subcontinent to South-Western China, India, Bangladesh, etc.

Occurrence

Occurs at deciduous forest, gardens and light scrub areas throughout year. Mostly occur in hilly regions of India.

Marks of Identification

It is about 7.8-9.0 inch and 1-2 inches in length and breadth respectively. It shows glossy black chin and throat and slightly fufted head. Back and breast feathers brownish with white tip. Tail is with brownish feathers. There is crimson patch below the roof of the tail.

Life Cycle

The bird appear to be monogamous. Male courtship display involves showing erect crimson under tail. Two broods are completed per season, male and female build nest and both may incubate within 10-14 days. Fledge 12 days.

Insect Preys

Small insects, flying insects, moths, caterpillars, winged termites.

Economic Importance

It is biocontrol agent of insect pests shown above.

6. Red Wattled Lapewing (Plate 23.1F)

Taxonomic Position

Class – Aves

Order – Charadriformes

Family – Charadriidae

Genus – *Vanellus*

Species – *indicus*

Distribution

It is commenst and most familiar in India.

Occurrence

Occurs at drying up beds of village tanks and sun baked fallow fields. It also occur near the ploughed fields and grazing grounds near damp places. Many times the bird cry even at the night.

Marks of Identification

It is a size of grey patridge. It is broze-brown above and white below with black breast, head and neck and crimson fleshy wasttle in front of eye. A broad white band from behind eyes run down the sides of neck to meets whitish under parts.

Life Cycle

Female lays about 3-4 eggs in the nest. The nest of this bird is merely an unlined depression or scrape in ground. Their call is all too known "Did ye do it ?" or "pity to do it". Their eggs normally have some shade of grey brown and blotched width blackish. Adults as well as newly hatched dowry chicks are perfect examples of natural camouflage.

Insect Preys

Small insects, aquatic insects, caterpillars, etc.

Economic Importance

Acts as biocontrol agent for above said insect pests.

7. Indian Robin (Plate 23.1G)

Taxonomic Position

    Class – Aves

    Order – Passeriformes

    Family – Muscicapidae

    Genus – *Saxicoloides*

    Species – *fulicata*

Distribution

India, Pakistan, Sri Lanka.

Occurrence

Found in scattered bushes and ground habitations, thatch roofs, road side hedges, stones, etc.

Marks of Identification

Indian Robin is 19 cm long including the long cocked tail. It has similar in shape to the smaller European Robin. Male is black with white wing patch and rusfty red under root of tail. Female is ashy brown with no wing patch and width paler chestnut from under side.

Life Cycle

The female lays about 5-6 eggs in the nest. The nest is prepared with dead leaves and mass lined with hairs and is big for the size of the bird. During a year two or some

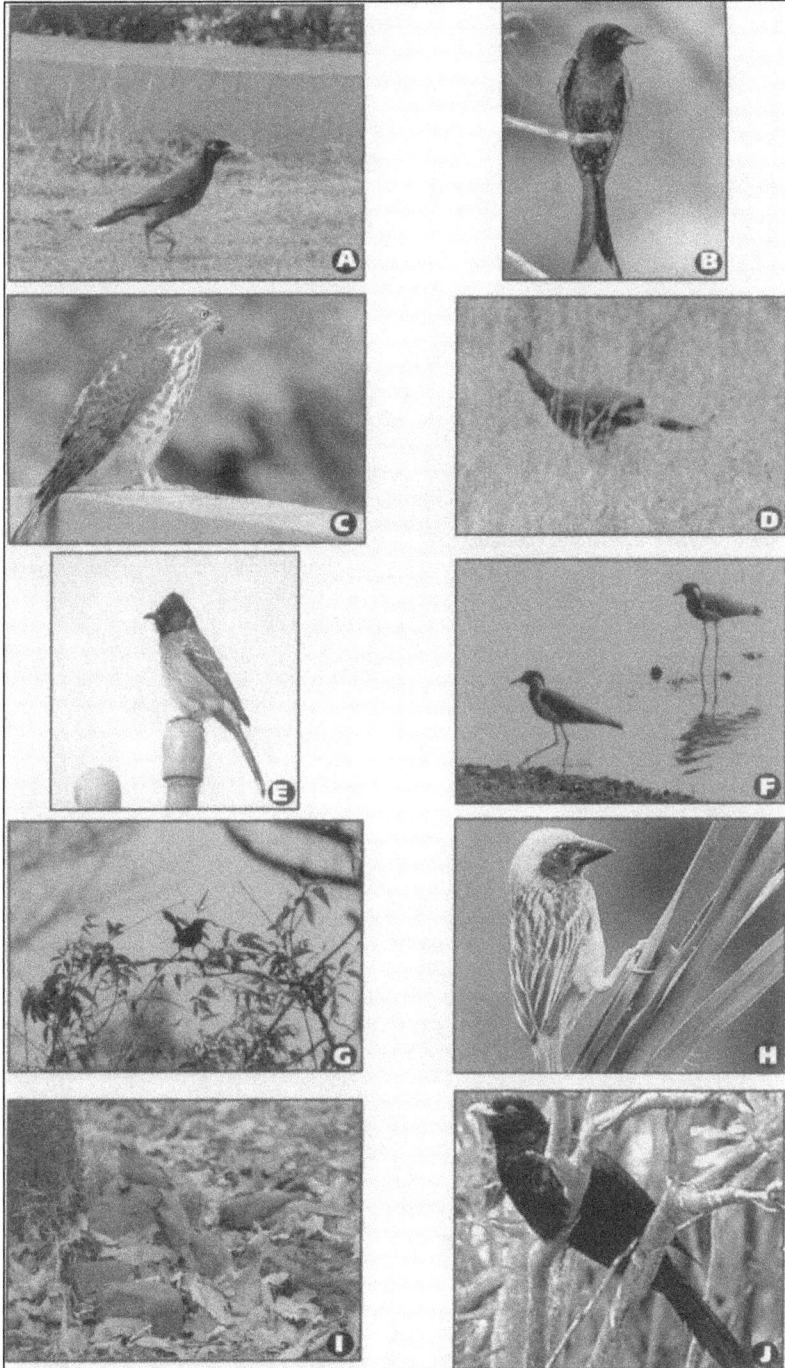

Plate 23.1: Birds of Biological Control of Insect Pests.

A: Indian Myna; B: Black Drongo; C: Shikra; D: Peacock; E: Red Vented Bulbul;
F: Lapewing; G: Indian Robin; H; Baya; I: Babbler; J: Koel.

Plate 23.2: Birds of Biological Control of Insect Pests.

K: Cattle Egret; L: Crow Pheasant; M: Pond Heron; N: Fowl; O: Green Bee Eater;
P: Hoopoe; Q: Crow; R: House Sparrow; S: Sunbird; T: Waterhen.

times three broods are completed. Nest building sometimes starts in March but often not till well on in April. Incubation and fledging each take about a fortnight. The young ones become ready to leave nest after wing formation.

Insect Preys

Robin's eat mainly insects especially caterpillars, termites, ants, crickets, grasshoppers, aphids, cicadas, etc.

Economic Importance

It is biocontrol agent of above harmful insects.

8. The Baya Weaver Bird (Plate 23.1H)

Taxonomic Position

Class – Aves

Order – Piciformes

Family – Ploceidae

Genus – *Ploceus*

Species – *phillippinus*

Distribution

Southeast Asia.

Occurrence

Baya forage in flocks for seeds, both on the plants and on the ground. They are known to glean paddy and other grain in harvested fields but occasionally damage ripening crops. They are abundant in Monsoon season and on paddy cultivation.

Marks of Identification

The Baya is sparrow sized 15 cm. In their non-breeding plumage, both males and females resembles with female house sparrow. They have a stout conical bill and a short square tail. Non breeding males and females look alike, dark brown streaked fulvous buff above, plain (unstreaked) whitish fulvous below. Eye brows of this bird are long and buff coloured. Bill is horn coloured and with no mask.

Life Cycle

After mating female lays about 2-4 white eggs and incubates them for about 17 days in the nest which is remarkably woven retort shaped. Young ones leave the nest after formation of wings upon them. The male prepares atleast 4-5 nests. The female can use these nests one after another. The nest is prepared with leaves of paddy and grasses.

Insect Preys

Butterflies, termites, small moths, etc.

Economic Importance

Acts as biocontrol agent of above said insects.

9. Babbler (Sat Bhai) (Plate 23.1I)

Taxonomic Position

Class – Aves

Order – Piciformes

Family – Muscicapidae

Subfamily – Timalinae

Genus – *Turdiodes*

Species – *caudatus*

Distribution

India and Sri Lanka.

Occurrence

The Babbler's habitat is forest and plains, agricultural fields. Often seen in gardens and found throughout the year.

Marks of Identification

The sexes are identical, drably coloured inbrownish grey. Young birds have dark iris. It has a long very rufous tail and dark primary flight feathers and has short expanded wings. It is a weak in flight. These birds always found in groups of seven hence, it is referred as "Sat Bhai".

Life Cycle

The eggs are turquoise blue. Female can lay about 3-4 eggs in the nest during June-September. Birds fledge and females tend to leave their natal group after 2 years. They are long lived and have been noted to live as long as 16.5 years.

Insects Preys

Grasshooppers, honey bee, spiders(non insesct), termites, caterpillars found on agriculture crops.

Economic Importance

Acts as biocontrol agent of above insects.

10. Koel (Plate 23.1J)

Taxonomic Position

Class – Aves

Order – Cuculiformes

Family – Cuculidae

Genus – *Eudynamys*

Species – *scolopacea*

Distribution

India, Sri Lanka and other Asian countries. In summer months the male start singing.

Occurrence

Found in gardens and groves. It is entirial arboreal never descends to the ground.

Marks of Identification

It is about size of the crow but slenderer with longer tail. Male is glistering with striking yellowish green bill and blood red eyes. Female is brown spotted barried with white. In the hot season the call of male is 'Kuoo-Kuoo-Kuoo'. It resides in counter side but female Koel has no song.

Life Cycle

Laying season corresponding with that of normal host house and Jungle Crow. Koel built no nest of its own but deposites its eggs in Crow nest leaving them to be hatched. Young to be reared by foster parents. Eggs are pale brownish but smaller than eggs of crow.

Insect Preys

Small beetles, hairy caterpillars, other small flying insects.

Economic Importance

It acts as biocontrol agent for above mentioned insect pests.

11. Cattle Egret (Plate 23.2K)

Taxonomic Position

Class – Aves

Order – Ciconiformes

Family – Ardeidae

Genus – *Bulbucus*

Species – *ibis*

Distribution

India, Sri Lanka, Myanmar, Thailand, China, etc.

Occurrence

Found in agricultural farms when ploughing is done, also seen with grazing cattle, ponds, ditches, etc.

Marks of Identification

It is white bird very similar to little egret but and always distinguished from in no-breeding season and by its stouter yellow bill. The bird shows orange buff or golden head and neck and back during its breeding plumage. Legs are long and black in colour.

Life Cycle

Breeding takes place during June-August in North India and November-March in South India. During breeding, colour of egret in golden colour. Mated female lays about 3-5 eggs of pale skim-blue in colour in nest and hatched young ones live in upto nest formation of wings upon them.

Insect Preys

Blue bottle flies, grasshoppers, cicadas, grubs, caterpillars, many insects parasites of cattles.

Economic Importance

The bird is also feeding on mosquito larvae, acts as biocontrol agent of above mentioned insect pests.

12. Crow Pheasant (Plate 23.2L)

Taxonomic Position

Class – Aves

Order – Cuculiformes

Family – Centropidae

Genus – *Centroopus*

Species – *sinensis*

Distribution

Worldwide. It is very common in India, Pakistan, Canada, China, etc.

Occurrence

The Crow Pheasant is dwellers of open scrub country abounding in bushes and small trees and cultivation or in grasses. It also found in neighborhood of human habitation.

Marks of Identification

Crow Pheasant measures about 48 cm. It is similar to forest Crow but have long tail and reddish yellowish wings. Body colouration is black. It shows reddish yellowish eyes.

Life Cycle

Mated female lay about 4 eggs in the nest. The pot shaped nest is prepared with leaves and branches of certain plants. It has breeding period from February to

September. The hatched young ones are developed in the nest upto the wing formation. The winged forms then leave the nest for individual feeding and mating purpose.

Insect Preys

Grasshoppers, caterpillars, some large insects, termites, ants, etc.

Economic Importance

Acts as biocontrol agent of above mentioned pests.

13. Pond Heron (Plate 23.2M)

Taxonomic Position

Class – Aves

Order – Ciconiiformes

Family – Ardeidae

Genus – *Ardeola*

Species – *groyii*

Distribution

Throughout India, Myanmar, Thailand, Malaysia, etc.

Occurrence

Found whenever there is water, river, jheel, Kutch well, temple, pond and roadside. It is found throughout the year.

Marks of Identification

It is smaller than the cattle egret. The Amarsh birds are chiefly earthy brown with pale brownish yellow head and neck. They also show glistening white wings, tail and rump flash into prominence when flys. It may seen singly or in loose parties. It stands hunched up and inert on the squelchy of the prey. The head is drawn in between the shoulders when the waits for prey.

Life Cycle

Mated female lays about 3-5 eggs in the nest. Eggs are pail greenish blue. The nest is Crow type. Its breeding season is chiefly from May to September. In breeding season the bird aquires maroon hair like plumes on back. Both sexes look similar.

Insect Preys

Mosquito larvae, midge larvae and other aquatic insects, etc.

Economic Importance

It acts as good biocontrol agents of various aquatic insect pests.

14. Red Jungle Fowl (Plate 23.2N)

Taxonomic Position

    Class – Aves

    Order – Galliformes

    Family – Phasianidae

    Genus – *Gallus*

    Species – *gallus*

    Subspecies – *domesticus*

Distribution

Widely distributed in Asiatic countries and also in UK, Ireland and Canada. It is widespread domestic bird in India.

Occurrence

It occurs chiefly in Himalayan Terrian and domesticated among human dwelling in India.

Marks of Identification

Male is darker, brighter colour than female (hen). Fowl is a sober mottled brown with some metallic green on neck and with red bill and vertically placed blackish or brownish tail like feathers.

Life Cycle

Mated female lays about 3 to 5 eggs. The young ones of fowl are chickens. The bird can live for 5 to 10 years depending upon the breeds. In commercial intensive farming, meat chickens generally lives only 14 weeks before slaughter. Hens of special laying breeds may produce 300 eggs per year. Their eggs are whitish and hatched within 21 days.

Insect Preys

It feeds on many insects such as caterpillars, grubs, termites, moths, beetles, ants, wireworms, cutworms, spodopterans, etc.

Economic Importance

The bird acts as biocontrol agent for pests mentioned above.

15. Small Green Bee Eater (Plate 23.2O)

Taxonomic Position

    Class – Aves

    Order – Coraciformes

    Family – Meropidae

Genus – *Merops*

Species – *orientalis*

Distribution

India, Africa, Southern Europe, Madagascar, New Guinea, etc.

Occurrence

Found throughout plains, on trees and bushes, etc. They are found diving differently and summer salting in air.

Marks of Identification

The bird measures about 21 cm. This grass green bird is tinged with reddish brown on head and neck. Central pair of tail feathers is prolonged. It shows slender, long, slightly curved bill. It found throughout the plains. Before eating insects bee eater removes the sting by repetedly hitting the insect on a hard surface.

Life Cycle

The mated female lays about 4-9 white eggs in the nest is prepared on any plant. It prepares one hole at the end of the nest. Bee eater is gregarious. They form colonies by nesting in burrows tunneled into the side of sandy bank such as those which have collapsed on the edges of rivers. This bird is monogamous and have biparental care of the young.

Insect Preys

Aphids, cicadas, mealybugs, many cell sap sucking insect pests, bees, small insects, etc.

Economic Importance

It is good predator of many insects in several agroecosystems and horticultural ecosystems.

16. Hoope (Plate 23.2P)

Taxonomic Position

Class – Aves

Order – Coraciformes

Family – Upupidae

Genus – *Upupa*

Species – *epops*

Distribution

Distributed throughout India. It has been reported from Maharashtra, M.P., U.P., Kerala, Tamil Nadu, etc.

Occurrence

It has been observed in plains and hills upto about 2000 m elevation. It is found with lawns, gardens and grows in and around villages and towns.

Marks of Identification

Hoope measures about 31 cm and is yellowish brown bird. It shows black and white strips on back, wings and tail. A conspicuous fan shaped crist and long, slender, curved bill is present. Feathers of crest are with white spots black tips. Sexes are similar to external look.

Life Cycle

Mated female can lays about 5-6 eggs. The eggs are white and laid in the nest. Its nest is in a hole in wall, roof or under of buildings or natural trees-hollow. The eggs become much soiled and discoloured during incubation. Nest is build during the period of February and May. The birds leave the nest after wing formation.

Insect Preys

Termites, bark beetles (Grubs and pupae), caterpillars, crickets, cockroaches, grasshoppers, etc.

Economic Importance

Its diets consists of insect, grubs and pupae. Hence it acts as good biocontrol agent of several agricultural pests.

17. Crow (Plate 23.2Q)

Taxonomic Position

Class – Aves

Order – Cucumiformes

Family – Corvidae

Genus – *Corvus*

Species – *splendens*

Distribution

Crows are found all over the world except in New Zealand, Antartica and South America.

Occurrence

They are mainely found in agricultural lands, farmland and wood lands throughout the year. They are also noticed around human habitations.

Marks of Identification

Adult Crow is black coloured with grey neck strip and little smaller than Jungle Crow. It measures about 45 cm from top of beak to end of tail and has a greenish blue glos on their wings.

Life Cycle

Eggs are pale blue green speckled and streaked with brown. The female can lay 4-5 eggs. Eggs are laid in untidy platform like nests. Baby Crow stay in the nest upto 2 months. When the wing developed, the young birds leave the nest with parents. Adult Crow can live upto 10 years.

Insect Preys

Crickets, locusts, termites, weevils, caterpillars including *Helicoverpa armigera*, *Spodoptera* spp., *Agrotis*, Semiloopers, etc.

Economic Importance

Acts as biocontrol agent for above pest insects.

18. House Sparrow (Plate 23.2R)

Taxonomic Position

Class – Aves

Order – Piciformes

Family – Ploceidae

Genus – *Passer*

Species – *domesticus*

Distribution

India, Pakistan, Himalaya region, etc.

Occurrence

Sparrow found in considerable number on farms, especially where grain is grown and they gather in flocks round the ricks and hedges.

Marks of Identification

The bird measures about 15 cm in size. It shows short beak, black eyes ringed with yellow and brownish black feathers. The ventral of neck is black coloured, dorsal of head and neck is light brown. Legs are grayish white. The wing and tail feathers are blackish to grey.

Life Cycle

Mated female bird lay about 3-5 eggs. The eggs are pale greenish motive and marked with brownish shade. The pair bound is long lasting during breeding and extended season but for the rest of the year, it is gregarious. In autumn they remaining in smaller flocks or group upto the onset of breeding. Breeding take place practically throughout the year. There are two or three broods in single year usually begins with May.

Insect Preys

It feeds on caterpillars, small flying insect pests like jassids, crickets, grasshoppers, termites, etc.

Economic Importance

It is associated with human dwellings. The bird is acting as a good biocontrol agents for insect pests on various agricultural crops.

19. Mountain Blue Bird (Plate 23.2S)

Taxonomic Position

Class – Aves

Order – Passeriformes

Family – Turdidae

Genus – *Sialia*

Species – *currucoides*

Distribution

The mountain blue bird is migratory. Northern birds which migrate to southern part of range. It occurs in western Asiatic countries, Canada, etc.

Occurrence

It is a migratory bird. Some birds may move to lower elevations in winter.

Marks of Identification

It is medium sized bird weighting about an ounce with length from 15 cm – 20 cm (6-8 inches). It is with light under belly and black eyes. Adult male have thin bills. They are bright blue and somewhat lighter beneath.

Life Cycle

Their breeding habitat is western Asia including mountain areas. The female bird prepare nest and after mating lay few eggs in the nest. The eggs are hatched by feamale by providing appropriate temperature to eggs on sitting upon them. Care of young ones is taken by female upto wing formation.

Insect Preys

Grasshoppers, dragonflies, locusts, some flying insects, termites, mosquitoes, midges, etc.

Economic Importance

Acts as a biocontrol agent for grasshoppers, locust, termites and midges, etc.

20. White Breasted Waterhen (Plate 23.2T)

Taxonomic Position

Class – Aves

Order – Ralliformes

Family – Rallidae

Genus – *Amaurornis*

Species – *phoenicurus*

Distribution

India, Sri Lanka, China, Indonesia, etc.

Occurrence

Occurs at neighborhood of water and casually met with singly or in pairs in pond region, ditches, etc.

Marks of Identification

It is common slaty grey stub-tailed, bare legged marsh bird. Adult is white breasted waterhen have mainly dark grey upper parts and flanks of white face, neck and breast. It shows long toes, short tail and yellow bill and legs.

Life Cycle

The bird is silent except in rainy season when it is breeding. The male become very pugndious noisy. The female lays about 6-7 eggs. Eggs are pinkish or cream or striked and blotched with reddish brown. The eggs are laid in nest and young ones cared upto the full formation of wing feathers.

Insect Preys

Some aquatic insects, small dipterous insects.

Economic Importance

The bird can act as biocontrol agent for aquatic insects like mosquitoes and gnats, etc.

Discussion

During study period, in all 20 birds have been recorded as insectivorous birds. Out of which Indian Mynah bird, Black Drongo, Green Bee Eater, Crow, Shikra found more potential. Indian Mynah bird *A. tristis* has tremendous potential to control grasshoppers. Indian Mynah bird was used in controlling red locust in Moritious in 1772. It was the first International movement of biological control of insect pest through birds. India shared the first International movement by sending Indian Mynah bird to Moritious. The birds listed in the text have good potential for utilization in biological pest control of insects. The birds may migrated from the target areas for their basic needs *i.e.* food, shelter and mate. Their migration may be checked by providing their basic needs and they may be utilized in pest control programmes. However, there is very scanty literature on utility of birds in insect pest control. The present work will add great relevance in designing biocontrol programmes for pest insects using birds.

Conclusion

On the basis of the present work it is concluded that Kolhapur region has very rich insect diversity. Hence, there are several kinds of insectivorous birds in the region. Making their index and assessing them as biocontrol agents is big task in future.

However, some birds should be reared on large scale and used in pests control. Indian mynah, king crow and cattle egret have tremendous potential in biocontrol programmes of insect pests.

References

Ali, Salim and Fatehally, Laeeq, 1967. *Common Birds.*

Bruns, H., 1957. The economic importance of birds in forests. *Bird Study*, 7: 192–208.

Chitampalli, Maruti, 2002. *Pokshikosh*, pp. 50.

Coppel, H.C. and Mertins, J.W., 1977. *Biological Insect Pest Suppression.* SUBH, New York, pp. 1–301.

Coppel, H.C. and Solan, N.F., 1971. Avian predation, an important adjuct in the suppression of larch caserbearer and introduced Pine sawfly population in Wisconsin forests. *Proc. Tall Timbers Conf. Ecol. Animal Contr. Habitat Management*, 2: 259–272.

George, J.L. and Mitchell, R.T., 1948. Calculations on the extent of spruce budworm control by insectivorous birds. *J. Forestry*, 46 : 454–455.

Gibb, J.A., 1960. Populations of tits and gold crests and their food supply in pine plantations. *Ibid*, 102: 163–208.

Hertig, B. and Simmonds, F.J., 1980. *A Catalogue of Parasites and Predators of Terrestrial Arthropods*, 1 : 1–122.

Knight, F.B., 1958. The effect of wood peckers on populations of tile Engelmann Spruce Bettle. *J. Econ. Entomol.*, 51: 603–607.

Krishnaswamy, N., Chauhan, O.P. and Das, R.K., 1984. Some common predators of the rice insect pests in Assam, India. *Int. Rice Res. Newsl.*, 9(2): 15–16.

Massey, C.L. and Wygant, N.D., 1944. Biology and control of the Engelmann Spruce Bettle in Colorado. *Proc. U.S. Dep. Agr. Ctre.*, pp. 944.

Nicholson, E.M., 1938. *Birds and Men: The Birds Life of Town, Villages, Garden and Farm Man.*

Sathe, T. V. and Bhosale, Y.A., 2001. *Insect Pest Predators.* Daya Publishing House, New Delhi, pp. 1–165.

Sathe, T.V., 2008. *A Textbook of Forest Entomology.* Daya Publishing House, New Delhi, pp. 1–234.

Sathe, T.V., Jadhav, B.V. and Kadam, S.D., 2008. Birds as pest biocontrol agents in agroecosystems of Satara district. *Biotechnological Approaches in Entomology*, 3: 170–177.

Steward, R.E. and Aldrich, J.W., 1951. Removal and repopulation of breeding birds in a sprucefir forest community. *Auk*, 68: 471–482.

Sweetman, H.L., 1958. *The Principle of Biological Control*. WmC. Brown, Dubuque.

Tinbergen, L., 1960. The natural control of insects in pinewoods 1. Factors influencing the intensity of predation by songbird. *Arch. Neerl. Zool.*, 13: 265–336.

Van vreden and Abdul, 1986. Pest of rice and their natural enemies in peninsular Malaysia, pp. 151. Pudoc Wageningen.

INDEX

www.ingramcontent.com/pod-product-compliance
Lightning Source LLC
Chambersburg PA
CBHW021435180326
41458CB00001B/280